Lake Tourism

ASPECTS OF TOURISM
Series Editors: Professor Chris Cooper, *University of Queensland, Australia*
Dr C. Michael Hall, *University of Otago, Dunedin, New Zealand*
Dr Dallen Timothy, *Arizona State University, Tempe, USA*

Aspects of Tourism is an innovative, multifaceted series which will comprise authoritative reference handbooks on global tourism regions, research volumes, texts and monographs. It is designed to provide readers with the latest thinking on tourism world-wide and in so doing will push back the frontiers of tourism knowledge. The series will also introduce a new generation of international tourism authors, writing on leading edge topics. The volumes will be readable and user-friendly, providing accessible sources for further research. The list will be underpinned by an annual authoritative tourism research volume. Books in the series will be commissioned that probe the relationship between tourism and cognate subject areas such as strategy, development, retailing, sport and environmental studies. The publisher and series editors welcome proposals from writers with projects on these topics.

ASPECTS OF TOURISM 32
Series Editors: Chris Cooper (*University of Queensland, Australia*),
C. Michael Hall (*University of Otago, New Zealand*)
and Dallen Timothy (*Arizona State University, USA*)

Lake Tourism
An Integrated Approach to Lacustrine Tourism Systems

Edited by
C. Michael Hall and Tuija Härkönen

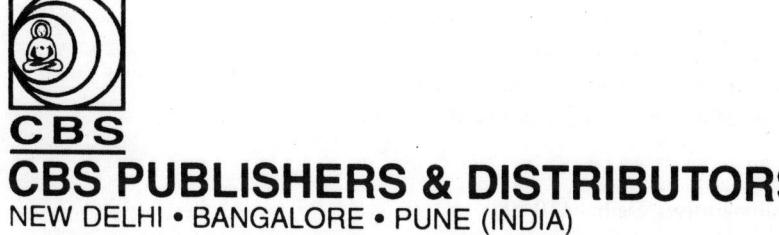

CBS
CBS PUBLISHERS & DISTRIBUTORS
NEW DELHI • BANGALORE • PUNE (INDIA)

CBS Publishers   ISBN : 978-81-239-1709-2
Channel View   ISBN : 978-1-84541-040-7

First Indian Reprint : 2009

Copyright © C. Michael Hall and Tuija Harkonen, 2006

This edition of Lake Tourism: An Integrated Approach to Lacustrine Tourism Systems is published by arrangement with Channel View Publications and is for sale only in the following territories: India, Sri Lanka, Bangladesh, Pakistan and Nepal only; not for export.

All rights reserved. No part of this book may be reproduced or transmitted in any form or by any means, electronic or mechanical, including photocopying, recording, or any information storage and retrieval system without permission, in writing, from the publisher.

Published by:
S.K. Jain and produced by V.K. Jain for CBS Publishers & Distributors,
4819/XI, 24 Ansari Road, Daryaganj, New Delhi - 110 002, India
e-mail: cbspubs@vsnl.com, cbspubs@airtelmail.in
Website: www.cbspd.com

Branches:
- ***Bangalore:*** 2975, 17th Cross, K.R. Road,
  Bansankari 2nd Stage, Bangalore - 560 070
  Fax: 080-26771680 • e-mail: cbsbng@dataone.in
- ***Pune:*** Shaan Brahmha Complex, Basement, Appa Balwant Chowk,
  Budhwar Peth, Next to Ratan Talkies, Pune - 411 002
  Fax: 020-24464059 • e-mail: pune@cbspd.com

Printed at :
Chaman Enterprises, Delhi - 110 095

Contents

List of Illustrations vii
List of Contributors x
Acknowledgements xii

Part 1: Introductory Context
1 Lake Tourism: An Introduction to Lacustrine Tourism Systems
   C. Michael Hall and Tuija Härkönen 3
2 Lakes as Tourism Destination Resources
   Chris Cooper 27

Part 2: Historical Dimensions and Cultural Values
3 Natural Heritage and the Maintenance of Iconic Stature: Crater Lake, Oregon, USA
   Stephen R. Mark 45
4 The Changing Historical Dimensions of Lake Tourism at Savonlinna: Savonlinna – The Pearl of Lake Saimaa. Lake Representations in the Tourist Marketing of Savonlinna
   Kati Pitkänen and Mia Vepsäläinen 67
5 Lakes, Myths and Legends: The Relationship Between Tourism and Cultural Values for Water in Aotearoa/New Zealand
   Anna Carr 83

Part 3: Tourist Activities and Perceptions
6 Lakes as an Opportunity for Tourism Marketing: In Search of the Spirit of the Lake
   Anja Tuohino 101
7 Lake Tourism in the Netherlands
   Martin Goossen 119

8 The Ostroda–Elblag Canal in Poland: The Past and Future for Water Tourism
Grazyna Furgala-Selezniow, Konrad Turkowski, Andrzej Nowak, Andrzej Skrzypczak, Andrzej Mamcarz 131
9 Finnish Boaters and Their Outdoor Activity Choices
Tuija Sievänen, Marjo Neuvonen and Eija Pouta 149

Part 4: Planning and Management Issues

10 Planning and Management of Lake Destination Development: Lake Gateways in Minnesota
William C. Gartner 167
11 Lake Tourism in New Zealand: Sustainable Management Issues
C. Michael Hall and Michelle Stoffels 182
12 Local Considerations in Marketing and Developing Lake-destination Areas
Daniel L. Erkkilä 207
13 Research Agendas and Issues in Lake Tourism: From Local to Global Concerns
C. Michael Hall and Tuija Härkönen 223

Index 234

List of Illustrations

Figures

1.1	The lacustrine system	13
1.2	Elements in integrated lacustrine management	15
4.1	Picture from a tourist brochure entitled 'Savonlinna à la Carte': St Olof's Castle, a steamboat and lake landscape	68
6.1	German respondents' comments on picture 'Cyclists'	110
6.2	Italian respondents' comments on picture 'Cyclists'	111
6.3	German respondents' comments on picture 'Angling'	112
6.4	Italian respondents' comments on picture 'Angling'	113
6.5	German respondents' comments on picture 'Midsummer Bonfire'.	114
6.6	Italian respondents' comments on picture 'Midsummer Bonfire'	115
7.1	Coastal and inland waterways of the Netherlands	120
7.2	Number of passages of pleasure craft counted in locks	122
7.3	Use of the waters by sailing boats	125
7.4	Use of the waters by motor craft	126
8.1	The Ostroda–Elblag Canal	133
9.1	Boating groups and groups of participants and non-participants in different recreational activities	155
9.2	Percentage of the Finnish population participating in rowing or small-craft motor-boating as a recreational activity by region	161
9.3	Percentage of the Finnish population participating in sailing or motor-cruising as a recreational activity by region	162
10.1	International tourism distribution channel	170
10.2	Planning horizon	174
10.3	Types of accommodation	175
10.4	Prior visitation	176
10.5	State of primary residence	177
12.1	Minnesota's marketing opportunity map	213

Tables

3.1	Annual visitation to Crater Lake National Park	48
3.2	US national parks located in the three west coast states, 2003	50
3.3	Visitation and funding for selected US national parks, 1910	58
3.4	Visitation and funding for selected US national parks, 1915	58
3.5	Size, visitation and budget for US national parks, 1934	60
4.1	Representations of Finnishness in the advertising material of Savonlinna	72
4.2	Representations of nostalgia in the advertising material of Savonlinna	74
4.3	Representations of the attractions of a 'holiday experience' in the advertising material of Savonlinna	76
4.4	Representations of environmental consciousness in the advertising material of Savonlinna	78
6.1	Responses divided into main categories and subcategories	109
7.1	Summary of water-related tourism in the Netherlands	121
7.2	Age (%) of skippers per type of boat, 1993 and 2002	123
7.3	Number of persons on board (%) per type of boat, 1993 and 2002	123
7.4	Dimensions per type of boat, 1993 and 2002	124
7.5	Median use of the boat per type of boat	128
7.6	Mean expenditure (€) of respondents during last 24 hours per type of boat	128
7.7	Percentage of respondents reporting encounters with specific problems during a boating holiday	129
8.1	Description of the lakes located in the main sections of the Ostroda–Elblag Canal	136
8.2	Accommodation facilities of the *powiats* (intermediate administrative divisions) of Elblag, Ilawa, Ostroda and the *voivodship* (province) of Warmia and Mazury in 2002	138
8.3	Accommodation facilities available in the *powiat* of Ostroda in 2004	139
8.4	Tourist flows on the Ostroda–Elblag Canal – the region of the slipways and the system of the Warmia lakes in 1996–2002	141
8.5	Annual tourist flows in the region of the locks and slipways in 1996–2002	142
8.6	Average seasonal distribution of tourist traffic in the region of the Slipways of the Ostroda–Elblag Canal in 1996–2002	143

9.1	Boating groups described according to choices of other recreational activities	156
9.2	Number of boating days or times, distance to boating site, number of vacation days and amount of money used for outdoor recreation in the household on average in different boating groups	158
9.3	Description of boaters according to their socioeconomic characteristics	159
10.1	Average daily expenditure per party	175
11.1	Lake- and lacustine-related activities and attractions undertaken by international visitors in New Zealand, 1997–2000	184
11.2	Ten largest lakes in New Zealand	186
11.3	Potential lakes of national importance for recreation, identified by the Ministry for the Environment	188
11.4	Phases in the development of a strategy to manage nutrient inflow into Lake Taupo	191
11.5	Goals of strategy for the Rotorua lakes	193
11.6	Objectives and principles for New Zealand's water Programme of Action	201
11.7	Summary of preferred package of actions with respect to fresh water in New Zealand	202
12.1	Underlying motives for taking a holiday trip	210
12.2	'Top of mind' associations with Minnesota	211
12.3	Minnesota's unique equities as a holiday destination	211
12.4	Visitor ratings of elements important in selecting the Itasca lakes area as a destination	214

Contributors

Anna Carr, Department of Tourism, University of Otago, PO Box 56, Dunedin, New Zealand. Email: acarr@business.otago.ac.nz

Chris Cooper, School of Tourism and Leisure Management, The University of Queensland, 11 Salisbury Road, Ipswich, Queensland 4305, Australia. Email: c.cooper@uq.edu.au

Daniel L. Erkkilä, University of Minnesota, 1861 E. Highway 169, Grand Rapids, Minnesota, USA. Email: erkkila@umn.edu

Grazyna Furgala-Selezniow, Department of Lake and River Fisheries, University of Warmia and Mazury in Olsztyn, ul. M. Oczapowskiego 5, 10–957 Olsztyn, Poland. Email: graszka@.uwm.edu.pl

Bill Gartner, Applied Economics, University of Minnesota, St. Paul, Minnesota, USA. Email: wcg@umn.edu

Martin Goossen, Alterra, Green World Research, PO Box 47, 6700 AC Wageningen, The Netherlands. Email: martin.goossen@wur.nl

C. Michael Hall, Department of Tourism, University of Otago, PO Box 56, Dunedin, New Zealand. Email: cmhall@business.otago.ac.nz

Tuija Härkönen, MBA Programme Coordinator, Research and Business Development Centre, Helsinki Business Polytechnic, Ratapihantie 13, 00520 Helsinki, Finland. Email: tuija.harkonen@helia.fi

Andrzej Mamcarz, Department of Lake and River Fisheries, University of Warmia and Mazury in Olsztyn, ul. M. Oczapowskiego 5, 10–957 Olsztyn, Poland.

Stephen R. Mark, US National Park Service, PO Box 3, Crater Lake, Oregon 97604, USA. Email: Steve_Mark@nps.gov

Marjo Neuvonen, Finnish Forest Research Institute, Unioninkatu 40 A, 00170 Helsinki, Finland. Email: marjo.neuvonen@metla.fi

Andrzej Nowak, Department of Geodesy and Land Management, University of Warmia and Mazury in Olsztyn, 10–957 Olsztyn, Poland.

Kati Pitkänen, Savonlinna Institute for Regional Development and Research, Kuninkaankartanonkatu 5, PL 126, 57101 Savonlinna, Finland. Email: kati.pitkanen@joensuu.fi

Eija Pouta, Agrifood Research Finland, Luutnantintie 13, 00410 Helsinki, Finland. Email: eija.pouta@mtt.fi

Tuija Sievänen, Finnish Forest Research Institute Unioninkatu 40 A, 00170 Helsinki, Finland. Email: tuija.sievanen@metla.fi

Andrzej Skrzypczak, Department of Lake and River Fisheries, University of Warmia and Mazury in Olsztyn, ul. M. Oczapowskiego 5, 10–957 Olsztyn, Poland.

Michelle Stoffels, Department of Tourism, School of Business, University of Otago, PO Box 56, Dunedin, New Zealand. Email: michelle.stoffels@otago.ac.nz

Anja Tuohino, Savonlinna Institute for Regional Development and Research, University of Joensuu, Kuninkaankartanonkatu 5, P.O. Box 126, 57101 Savonlinna, Finland. Email: anja.tuohino@joensuu.fi

Konrad Turkowski, Department of Lake and River Fisheries, University of Warmia and Mazury in Olsztyn, ul. M. Oczapowskiego 5, 10–957 Olsztyn, Poland.

Mia Vepsäläinen, Savonlinna Institute for Regional Development and Research, Kuninkaankartanonkatu 5, PL 126, 57101 Savonlinna, Finland. Email: mia.vepsalainen@joensuu.fi

Acknowledgements

Although lakes have long been a focal point for leisure and tourism activity, it is only in recent times that they have become a foci of tourism research. The first international academic conference on lake tourism was held in Savonlinna in Finland in 2003, the second in China in 2005. Indeed, concerns over environmental quality, amenity migration, regional development, access to fresh water and environmental change mean that tourism has now become a central issue for those interested in the management and sustainability of lakes and lacustrine systems. Nevertheless, it must also be remembered that lakes are often locations of intense personal and place attachment for people who live on them or visit them. In Nordic countries, for example, there are arguably few locations like a quiet forested lake to make one contemplate a sense of belonging or feelings of absence and longing – the generation of a positive response of wishing you were there. Indeed, in the Nordic context, it is readily apparent that the landscape and water in particular, plays a strong part in senses of national and regional identity, a feeling that is also shared in many other regions of the world.

It is within this context that this volume has been developed. Its origins lie within the Lake Tourism Project that was developed within the Finnish University Network of Tourism Studies (FUNTS) and managed by the Savonlinna Institute for Regional Development and Research at the University of Joensuu. That lake tourism is associated with Finland and Savonlinna in particular should come as no surprise for those with an interest in the area. There are arguably few countries that are so dependent on lakes as an image and activity factor in their tourism than Finland, while Savonlinna is a destination whose development is inseparable from that of its lake setting. However, as this volume demonstrates, while the Finnish experience with lake tourism provides a significant source of knowledge on the subject, interest in lake-related tourism is now a worldwide phenomena.

In addition to the support provided by the Finnish University Network of Tourism Studies and the Savonlinna Institute for Regional

Development and Research, the editors would particularly like to acknowledge the role of Arvo Peltonen, Pellervo Kokkonen and Hannu Ryhänen in developing Finnish academic interest in lake tourism. Michael would also like to thank especially Esa Ahola, Thor Flognfeldt, Stefan Gössling, Tuija, Dieter Müller, Stephen Page, Jarkko Saarinen, Anna Dora Saethorsdottir, Sandra Wall and Brian Wheeler for their support, unwitting or not, of his lake tourism research, as well as the background influence of Fiona Apple, Jackson Browne, Nick Cave, Bruce Cockburn, Elvis Costello, Stephen Cummings, Lucinda Wilson and Chris Wilson. David Duval also read drafts of parts of the manuscript. Thanks must also be given to Jody Cowper for her invaluable support, as well as the assistance of Monica Gilmour with getting the final manuscript printed and posted and, of course, to all the staff at Channel View Publications.

C. Michael Hall, *City Rise*
Tuija Härkönen, *Savonlinna*

Part 1: Introductory Context

Chapter 1
Lake Tourism: An Introduction to Lacustrine Tourism Systems

C. MICHAEL HALL AND TUIJA HÄRKÖNEN

In many parts of the world lakes are a vital part of recreation and tourism as both a location for leisure activities, as well as an attraction in their own right. Lakes are also used extensively by many countries and destinations in tourism promotion campaigns, whether it is to provide a key image of the destination or an attractive backdrop for other leisure activities (Härkönen 2003). Yet despite the general attention given to the role of the natural environment in tourism and recreation, including the maritime environment, remarkably little research has been undertaken on the role of lakes in tourism, although the impacts of tourism and recreation on the freshwater environment have been long recognised (e.g. King & Arnett 1974; Liddle & Scorgie 1980; Edington & Edington 1986; Newsome et al. 2002). Arguably, this attention given to lakes is not isolated to tourism studies alone. Odada (2004) bemoans the lack of basic scientific research on the world's freshwater lakes in comparison to that which is undertaken on the world's oceans, noting that 'we know far less about the biology and physics of Lake Victoria, for example, than we know about these aspects of the Indian Ocean' even though 'several characteristics of large lakes make them ideal study sites to advance our understanding of ecosystem dynamics' (2004: 161).

Despite their environmental significance, lakes are regarded as being grossly under-represented as ecosystems in the world's system of national parks and reserves (Vreugdenhil et al. 2003). Lake systems are the least protected of the world's biomes, biome being a major ecological community of plants and animals extending over a large natural area that has evolved so that they are adapted to the range of environmental factors, such as climate, within them. The International Union for the Conservation of Nature and Natural Resources (IUCN) use the areal classification of Udvardy (1975) in their determination of the world's biomes. In 2003 only 1.54% of the global lake system biome was included in a protected area (some 3,083 sq. miles (7,989 km^2) out of a possible 199,830 sq.

miles (517,695 km²)) (Chape *et al.* 2003). The IUCN defines a protected area as 'An area of land and/or sea especially dedicated to the protection and maintenance of biological diversity and of natural and associated cultural resources and managed through legal or other effective means' (Chape *et al.* 2003: 2). Some 261 protected area sites in lake system biomes were recorded, meaning that the average area of sites was only 12 sq. miles (30.6 km²). In addition, lakes – even those with a degree of legislative protection – are facing increasing environmental stress as a result of land-use change, population growth and pollution within the watershed that drains into them. It is therefore perhaps not surprising that 'water issues', of which lakes are clearly a part, 'have gradually made it to the top of our consciousness about environmental issues' (Adeel 2003: 1). Although it is also equally likely that many people in the developed world, particularly those who are involved with tourism, did not realise that 2003 was the International Year of Freshwater.

This volume therefore seeks to make a contribution to our understanding of the relationship between lakes and tourism. The focus is primarily on the developed world, given that is where the majority of the world's tourism occurs, although that is not to ignore the role that lakes play in the tourism industries of many developing countries, including the alpine lakes of the Andes (Carabelli 2001), such as Lake Titicaca; the Great Rift lakes in Africa, especially with respect to nature-based tourism (Zyl *et al.* 2000; Kassilly 2003); and various lake-based destinations in Asia such as Srinigar in Kashmir or Lake Toba in Indonesia (Diniyati 2002). The chapters are broadly organised into three main sections: Chapters 3–5 deal with historical dimensions and cultural values, Chapters 6–9 discuss tourist activities and perceptions with respect to lake and water-based tourism and recreation, while chapters 10–12 examine some of the planning and management issues associated with lake tourism. This and the following chapter discuss the role of lakes as destination resources, as well as some of the major issues that are associated with lake tourism. This chapter discusses the significance of lake tourism, environmental issues associated with lakes and the development of institutional arrangements for lake management that incorporate tourism concerns.

The Significance of Lake Tourism

As with many types of tourism, the economic and tourism significance of lakes is difficult to determine. In part, this is an issue of definition as lake tourism is tourism that occurs not only on the lake itself, but also in the surrounding area. Lacustrine tourism systems therefore include the lake, the foreshore and those amenities, facilities and infrastructure in the surrounding region that support the role of the lake as a tourist attraction. The environmental system that underlies the lake tourism system is usually much larger in area and includes all areas of the watershed that feed into the lake. Indeed, the idea of lake tourism reinforces the idea that there are certain geographical entities that, because of their particular environmental characteristics are often designated as a separate type of tourism in which the specific environment serves to attract particular activities and which serve to convey certain environmental images as part of destination promotion. Therefore, lake tourism can be distinguished and therefore understood, in much the same way that alpine or forest tourism have been recognised as a subfield of tourism studies.

Because of the difficulty in definition, there are also no separate set of statistics that solely detail how many people visit lakes for recreation and tourism purposes. This situation also means that there is no comprehensive analysis of the economic significance of lake tourism, although studies have been undertaken of particular types of lake-based recreational activities (e.g. Fadali & Shaw 1998), such as fishing which is recognised as one of the most significant activities (Jakus *et al.* 1998; Eiswerth *et al.* 2000; Lee 2001, 2002, 2003a, b; Provencher *et al.* 2002; Henderson *et al.* 2003), pilgrimage (Maharana *et al.* 2000), as well as research on the role of lakes in nature-based tourism (e.g. Yang *et al.* 1997; Zhang 1999; Yan *et al.* 2000; Vogelsong & Ellis 2001). For example, Lee (2002) identified that owners of 652,000 registered boats logged 18.4 million days of boating and spent US $635 million on trips within Michigan in 1998. All northern regions in the state were net gainers from boat-trip spending and earned a net gain of US$120 million in 1998. The south inland region of the state showed the biggest net loss of boater dollars, as resident boaters in this region spent US$78 million more outside the region than what the region received. Out-of-state boaters spent US$35 million in Michigan, mostly involving the use of seasonal homes in the state.

Yet regardless of the absence of a comprehensive and systematic account of lake tourism on a global scale there are clearly a number of destinations for which a lake or lake system is critical for their tourist attractiveness. There are many lake-based destinations around the world for which visitation runs into the millions – for example, the lake districts of England (Leslie 2002), Finland (see Vepsäläinen & Pitkänen, Chapter 4; Tuohino, Chapter 6) and Minnesota (see Gartner, Chapter 10), as well as the lake systems of Hungary, Scotland, Switzerland, northern Italy, New Zealand (see Carr, Chapter 5; Hall & Stoffels, Chapter 11), western Canada and the United States (Castle 2003; see Mark, Chapter 3) as well as, of course, the Great Lakes of North America, which are arguably the most researched lake systems in the world in terms of tourism and recreation (e.g. Connelly *et al.* 1999; Lee 2001, 2002, 2003a, b; Lockwood, Peck & Oelfke 2001; Mangun & O'Leary 2001; Payne *et al.* 2001).

Yet it is important to stress that while at present lakes are regarded in the present-day as attractive locations in Western countries and are focal points not only for tourism and recreation but also for amenity migration, this has not always been the case. Arguably, it is only since the late 18th century and the development of the Romantic movement that lakes came to be seen as attractive locations for leisure. Until this time lakes were generally seen in purely utilitarian terms as food and water sources. Indeed, to an extent lakes were often seen, as many wild lands, as being dangerous places because of the extent to which they harboured water-borne disease. However, the Romantic movement substantially changed perspectives on lakes and lake environments and as the West became more industrialised, so lakes and the landscape in which they were situated came to be seen more favourably (Buzard 1983). For example, England's most visited national park, the Lake District National Park (Lake District National Park Authority 1999) can trace its tourist popularity in terms of the attractiveness of its landscape to its popularization by the Lake District poets, William Wordsworth and Samuel Taylor Coleridge, at the beginning of the 19th century (Walton & McGloin 1981; Brennan 1987). It should be noted, however, that it did not take very long for tourism to be criticised as being damaging to that same district. For example, in 1848 Thomas Cook wrote in his handbook for visitors to Belvoir Castle in Leicestershire:

> It is very seldom indeed that the privileges extended to visitors of the mansions of the nobility are abused; but to the shame of some rude folk from Lincolnshire, there have been just causes of complaint at Belvoir Castle: some large parties have behaved indecorously and they have to some extent prejudiced the visits of other large companies. Conduct of this sort is abominable and cannot be too strongly reprobated. (Ousby 1990: 89)

Another important element in the development of the Romantic appreciation of lake landscapes is the extent to which they contributed to the development of national identity and senses of cultural independence. For example, in the latter half of the 19th century the Finnish lake landscape became commodified by the intelligentsia as an essential element of national cultural identity that served to contribute to Finnish independence from Russia (see Vepsäläinen & Pitkänen, Chapter 4). Similarly, indigenous expressions of nature in the United States, such as mountains, lakes, waterfalls, rivers and forests, were regarded as part of America's cultural declaration of independence from Britain and Europe (Kline 1970; also see Mark, this volume, Chapter 3). Nevertheless, although lakes became an important component of national identity and hence an element in the development of the national park ideal, it should be noted that there is substantial commonality within Western countries regarding the extent to which freshwater lakes formed a significant part of the landscape ideal from the mid-19th century on. Indeed, many of the present-day lake resorts in Europe, North America and New Zealand can date their origins back to this period. Nevertheless, while lake tourism has a history of 200 years, it is only in recent years that tourist pressure on lakes has started to cause major concern (e.g. Puczkó & Rátz 2000). However, it must be emphasised that the environmental stress caused by tourism is only one source of impact on lake quality, with upstream changes in land-use, increased urbanisation, poor sewage infrastructure and other sources of pollution and over-use of lake resources, all affecting lake quality.

Lake Environments and Tourism

The impact of tourism on lake environments is a function of the type of activities being engaged in, the number of people engaged and the nature of the lake environment itself. Although lake tourism is often considered as a rural activity (e.g. see Gartner, Chapter 10), it should be emphasised that in a number of countries urban lakes, including human-made lakes and reservoirs, constitute an extremely important recreational and tourism resource. For example, Hickley et al. (2004) noted the significance of freshwater angling as a recreational activity in Britain and described an initiative of the Environment Agency in England and Wales to identify lakes suitable for rehabilitation and develop an urban fishery development programme. In addition, there is often substantial pressure on lakes in the periurban areas of large urban conurbations and those within day-tripping distance for urban populations (e.g. Curtis 2003; Pickles 2003). In many developed countries lakes are experiencing substantial second- and retirement-home development as part of a broader pattern of amenity-related migration and human mobility that means that many holiday communities are gradually being transformed to substantial urban settlements through the processes of tourism urbanisation (e.g. Truly 2002; Marcouiller et al. 2004). Of course, in such situations the lakes that serve as the focal point of the environmental values are then often placed under increased pressure because of the changes in the water catchment for example, because of deforestation (Haigh et al. 2004) or other land-use practices (Hillman et al. 2003) and consequent inputs into the lake system such as pollution or increased sedimentation. Where these processes do occur, tourism-related activities both contribute to and are affected by environmental change in the lacustrine system (Puczkó & Rátz 2000; Hadwen et al. 2003; Mosisch & Arthington 2004; Jones et al. 2005). For example, in the case of the perched lakes of Fraser Island, Queensland, Australia, a study by Hadwen & Arthington (2003) indicated that clear lakes were the preferred swimming sites for 70% of respondents (n = 154), yet there is increasing concern that unregulated tourist activities may threaten the long-term conservation of the lakes.

Water-borne tourism activities such as boating and fishing can impact the environment through their effect on habitats and species and the potential impact on water quality via pollution from water craft as well as impacts on lake-edge vegetation (Zyll de Jong et al. 2000; Madrzejowska & Borowski 2003; Steiner & Parz-Gollner 2003; Mosisch & Arthington 2004). In many locations the impacts of water-based recreation has necessitated the development of extensive habitat protection programmes (e.g. Stauber 2004). Nevertheless, it cannot be assumed that all lake tourism and recreational impact is negative for species. For example, with respect to the New Zealand dabchick (*Poliocephalus rufopectus*), a protected endemic New Zealand grebe, confined to the North Island mainland and classified as vulnerable, Bright et al. (2004) found that human-made structures (jetties and houses) and recreational activities (boating) do not significantly affect the numbers and distribution of New Zealand dabchick pairs or nests in the bays of lakes Rotoiti, Tarawera and Okareka. Furthermore, the number of man-made structures was positively correlated with the number of chicks in the sampled bays. Bright *et al.* (2004) suggest that humans and dabchicks may be distributed similarly around the lakes because factors such as wind exposure and shoreline topography made certain sites preferable for both species. Alternatively, man-made structures may provide protected nesting environments and/or cover for chicks from predators, refuges from harassment by other bird species, or other benefits.

Arguably, in the vast majority of lake tourism destinations the greatest impact on the lake environment comes from non-tourism sources (Karakoc et al. 2003). For example, Lake Kinneret, in the north of Israel, is the only freshwater body in the country and is significant for recreation, tourism, a commercial fishing industry and as a water supply for other parts of the country. Consequently, maintaining a high water quality of the lake is of extremely high importance. The major part (some 90%) of the annual run-off of water enters Lake Kinneret from the north via the River Jordan during the autumn–winter floods. However, during this period, the river carries sediments, toxic agricultural chemicals and allochthonous organisms, including pathogenic bacteria, into the lake, thereby potentially damaging the lake's water quality and environment (Wynne et al. 2004).

Pollution may also have occurred in some lakes for a substantial period of time. For example, Lake Mälaren is the water supply and recreation area for more than one million people in central Sweden and is the subject of considerable environmental concern over heavy metal pollution. However, research indicated that the cause of the pollution was the 19th century and earlier extensive metal production and processing in the catchment, with the lake now experiencing a substantial improvement with respect to lead pollution, following closure of the mining and metal industry (Renberg *et al.* 2001). With respect to pollution issues in lake systems, it is important to recognise that arguably most pollution occurs not from activities on the water itself, but from those in the lake's water catchment, whether they are on the lake shore or upstream in source rivers of the lake (e.g. see Lee *et al.* 2002; Tambuk-Giljanovic 2003). Such a situation therefore highlights the necessity for taking a whole-of-catchment approach in ensuring the maintenance of water quality in lakes and, ideally, for wider management and development practices (Guo *et al.* 2001).

The effect of pollution can have a major impact not only on species living in the lake, but also on other animals further up the food chain, including human beings. Poor water quality can also have a substantial impact on tourism and recreational behaviour, either by influencing individual perceptions and behaviours or by direct regulation of access and recreational behaviour. For example, a study by Egan *et al.* (2004) on the use and value Iowans place on water quality in Iowa lakes found that approximately 62% of Iowa households visited one of the 130 lakes listed in the survey and the average number of trips per year was just over eight in 2002. Significantly, water quality was regarded as more important than either proximity or local park features in determining where households engaged in lake recreation.

There is a substantial body of literature surrounding the health impacts of eating polluted fish in the Great Lakes region of North America, often with a concentration on the angling community (e.g. Buck *et al.* 1999; Courval *et al.* 1999; Falk *et al.* 1999; Buck *et al.* 2000; He *et al.* 2001; Beehler *et al.* 2002; Cole *et al.* 2004; Dellinger 2004). In Wisconsin, consumption of Great Lakes fish is an important source of exposure to polychlorinated biphenyls (PCBs), dichlorodiphenyltrichloroethane (DDT), polybrominated diphenyl ethers (PBDEs) and other halogenated hydrocarbons, all of which may act as potential risk factors for breast cancer (McElroy *et al.* 2004).

The extent of heavy metal contamination in North American freshwater fish stocks is substantial. In 2002, in the United States, 48 states had issued advisories for sport-fish consumers that included 39 chemical contaminants. The most commonly identified chemical was methyl mercury, which is linked to reproductive and developmental effects. However, advisories to reduce consumption of contaminated fish have been issued by states since the early 1970s (Anderson *et al.* 2004). In a survey of fish consumption in women of child-bearing age (aged 18–45) undertaken in 1998 and 1999 (n = 3015) Anderson *et al.* (2004) reported that 29% reported sport-fish consumption during the previous 12 months. Most women (71%) were aware of mercury's toxicity to a developing child (87% among those aware of an advisory and 67% among those unaware of an advisory). However, awareness of state advisories was only 20%, ranging by state from 8% to 32%. Women who were older, had more than a high-school education and had a household member with a fishing licence were the most informed about mercury and fish-consumption advisories. In contrast, a 1998 survey of over 800 single-day beach visitors to Lake Erie's shoreline in Ohio found that nearly two-thirds of visitors take advantage of this information when making decisions about beach trips (Murray et al. 2001).

Another significant factor associated with water quality is the potential for lake users to catch waterborne diseases (Levy et al. 1998; Paunio et al. 1999). Yoder et al. (2004) reported on waterborne disease outbreaks (WBDOs) associated with recreational water in the USA, during January 2001–December 2002. During this period, 65 WBDOs associated with recreational water were reported by 23 states. These 65 outbreaks caused illness among an estimated 2536 persons, of which 61 were hospitalised and eight died. According to Yoder et al. (2004), this was the largest number of recreational water-associated outbreaks to occur since reporting began in 1978. Of these, 65 outbreaks, 30 (46.2%) involved gastroenteritis. These outbreaks of gastroenteritis were associated most frequently with *Cryptosporidium* (50%) in treated water venues and with toxigenic *Escherichia coli* (*E. coli*) (25%) and Norovirus (25%) in freshwater venues. Eight (12.3%) of the 65 recreational water-associated disease outbreaks were attributed to single cases of primary amoebic meningoencephalitis caused by *Naegleria fowleri*; all 8 cases were fatal and were associated with swimming in a lake (n = 7) or river. The 21 outbreaks involving dermatitis and four involved acute respiratory illness were associated with spas and pools. Although this was a record reporting period in terms of WBDOs, the authors of the report were unable to state whether the increase in disease outbreaks was a result of worsening water quality, improved surveillance and reporting at the local and state level, a true increase in the number of WBDOs, or a combination of these factors (Yoder et al. 2004). Nevertheless, in many lakes there is clear evidence of worsening water quality as a result of land-use change and practices in the catchment, particularly with respect to diseases such as *E. coli* (Chamot et al. 1998; Feldman et al. 2002; Bruce et al. 2003; McLellan & Salmore 2003) and leptospirosis (Morgan et al. 2002). For example, Jin et al. (2003, 2004) reported on the inflow of *E. coli*, enterococci and faecal coliform into Lake Pontchartrain (Louisiana, USA) as a result of stormwater run-off from the increasingly urbanised and drained catchment. Indeed, the largest outbreak of leptospirosis that has been reported in the USA occurred following a triathlon that had been held at Lake Springfield, Springfield, Illinois, after heavy rains (Morgan et al. 2002).

Although the interrelationships between tourism and environmental change in lakes are clearly significant, it must be emphasised that, for most lakes, tourism and recreation are only one part of the human dimensions of environment change. Indeed, in many cases other human dimensions of change such as agriculture, industrial development, urbanisation and other land-use practices will have a far greater impact than tourism. It is arguably only in lake systems, including their watersheds, that occur in wilderness areas or in other environments with high natural values, that tourism and recreation become the focal point of management and the key variable to be managed with respect to lake quality. In such cases there is a significant body of literature that can be drawn upon with respect to management strategies and practices that seek to balance the satisfaction of different user groups with the maintenance of environmental values (e.g. Christensen & Cole 2000; Kuentzel & Heberlein 2003). However, apart from lakes in some of the more peripheral areas of the world or those completely contained within national parks and reserves, the reality is that most lacustrine systems have substantial amounts of non-tourism activity within their boundaries. This situation therefore means that tourism and recreation, rather than being the focal point of management, becomes just one component in a complex web of factors that influence the lake environment and social and economic well-being. The wide range of factors that influence lake quality and stakeholders, interested in lake quality as well as water resources means that increasingly attention is focused on the development of collaborative or ecosystem management strategies that seek to integrate interests and issues in a more comprehensive approach to lake management. It is to this approach that we will now turn.

Integrated Lacustrine Management

Lacustrine systems are complex structures. The condition of any lake is dependent on upsteam inputs into the lake, as well as on factors in the lake itself. In the case of many lakes changed in land-use practices a great distance away, substantial impacts on the environment can follow for example, with respect to deforestation, agricultural practices and the discharge of industrial and human waste into rivers and streams that flow into a lake. Therefore, the effective management of any lake requires the integration of management over the entire catchment of the lake, as well as the lake itself. For example, in the case of Lake Biwa, the largest lake in Japan, that provides water to a population of 14 million people in Shiga, Osaka, Kyoto and Hyogo Prefectures, Yamamoto & Nakamura (2004) note that urban sprawl has resulted in serious land-use problems that are seriously aggravated in the south-eastern and north-eastern parts of the watershed and the expansion of urban areas has resulted in land-use control becoming more problematic, with consequent implications for lake quality.

Figure 1.1 indicates some of the key dimensions on a lacustrine system. A critical factor is the nature of land-use in the lake catchment area. Therefore, upstream agricultural and industrial practices can have an enormous impact on lake quality, often far greater than that of the immediate lake users. Another key issue in the management of lacustrine systems is that administrative and institutional boundaries do not neatly coincide with that of the lake watershed. This creates substantial difficulties in getting agreement between different governmental agencies and organisations with different mandates and jurisdictions. In the case of watershed management, a critical issue, of course, is why an upstream jurisdiction should be concerned with the impacts that they have on downstream jurisdictions. Indeed, this issue is fundamental to watershed management. In addition, destination region and promotional boundaries may also not easily coincide with that of a lake, even though the lake may be a key feature in tourism promotion. Indeed, a critical relationship issue here that often exists in lake tourism is that the agency responsible for the promotion of lake destinations is typically not the same agency that is responsible for environmental management, with consequent possible effects in terms of exceeding the environmental and social carrying-capacities of lakes. Such a situation also points to the fact that most lacustrine systems are subject to a range of objectives with respect to water use and quality that may at times even be conflicting (e.g. Yamamoto & Nakamura 2004). In many lacustrine systems around the world, one of the greatest problems is increased nutrient inputs into lakes and waterways, resulting in changes to biodiversity (some of which, for example, increase in the fish mass of target species may even be deemed desirable by some stakeholders), macro-algal blooms that can be dangerous to human health and eutrophication, the latter two impacts clearly being detrimental to tourism because of the perceived and actual damage to lake health. Invariably, the sources for increased nutrient input into lake systems comes from upstream catchment sources and not the immediate activities on the lake itself (e.g. Mulqueen 2002; Magadza 2003). Such a situation clearly highlights the need for a comprehensive approach to lacustrine management that links the various elements of the lacustrine system into an integrated management strategy.

'Waterways supply us with food and drinking water, irrigation for agriculture and water for aquaculture and horticulture. They are valuable assets for tourism and are prized recreational areas' (Department of the Environment 2003). It is the complexity of lacustrine systems that is demanding the development of integrated approaches to the management of lake systems. Figure 1.2 identifies key elements in integrated

Introduction to Lacustrine Tourism Systems

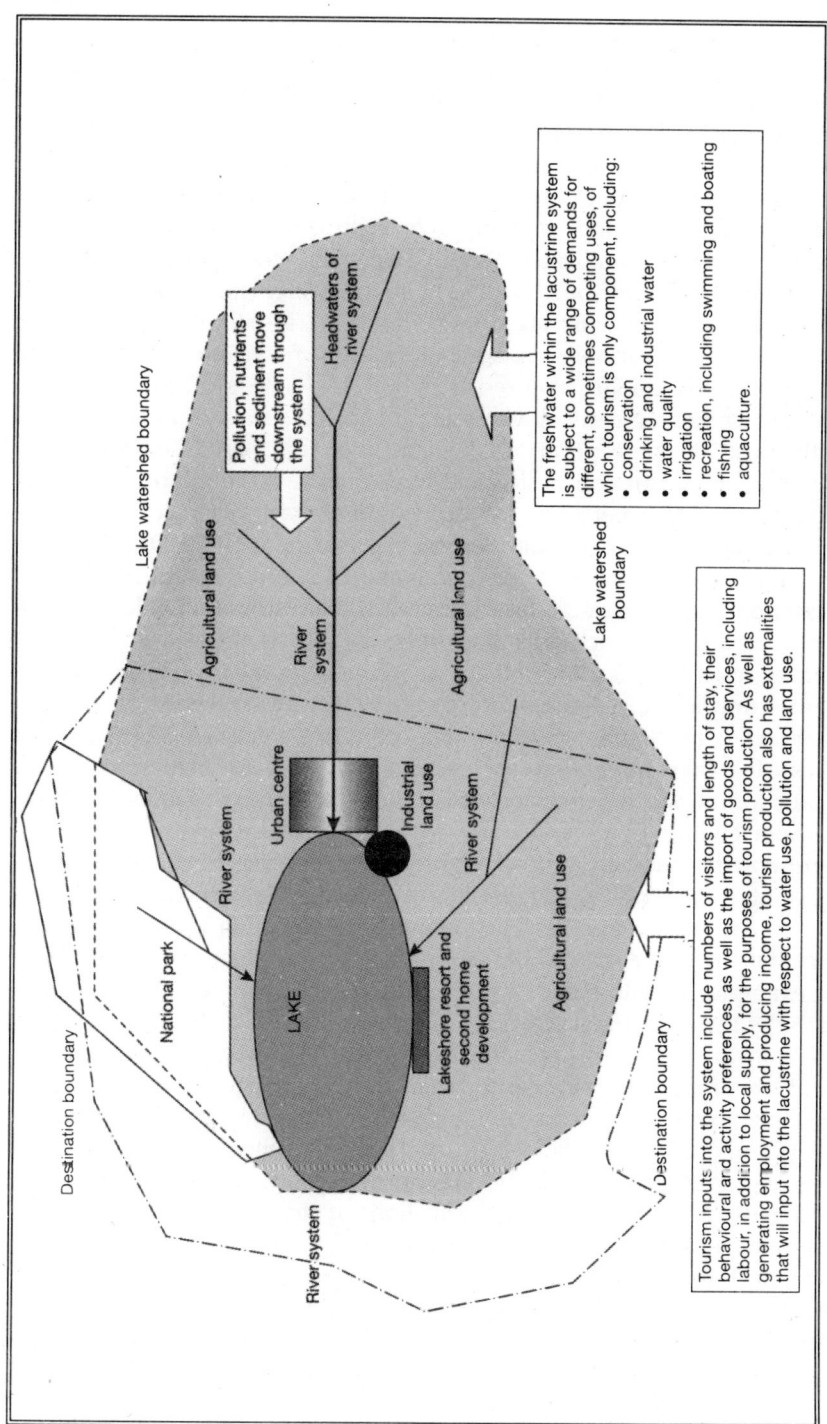

Figure 1.1 The lacustrine system

lacustrine management. Although bearing substantial similarities to catchment, watershed or integrated water resource management (Mitchell 1990; Rabe 1996; Heathcote 1998; Margerum 1999), integrated lacustrine management is different because the lake or lake system is of such central economic, social and environmental importance that it becomes the focal point of resource management strategies. Obviously, where lakes are not of such significance in their own right, they will become part of broader watershed and catchment management strategies.

A number of significant biotic and abiotic water quality variables can be identified for specific lakes and Figure 1.2 identifies some of these variables. Such variables become a means of empirically evaluating changes in lake quality as a result of both lake use and activities in the wider catchment. Nearly all variables are objective and although changes in the aesthetic perception of lakes and their associated landscapes are subjective they can still be tracked over time (e.g. Tuohino & Pitkänen 2004; also see Chapters 3 and 4, this volume). Historical accounts of the development, promotion and perceptions of lakes (e.g. Dredge 2001; Szylvian 2004; Carr, this volume, Chapter 5) may be useful in terms of corresponding changes in lake use and human impact, while information regarding lake activities and uses, as well as profiles of the users themselves, is also a key element in the development of lake management strategies (e.g. Reed & Parsons 1999; Reed-Andersen et al. 2000; Chapters 6 to 10 this volume). The objective variables identified for measuring lake quality are of great significance as they provide the means for measuring environmental change and the consequent changes in water quality, biodiversity and capacity to use the lake and therefore highlight the need to address significant problems. However, there is substantial evidence to suggest that it is extremely difficult to translate scientific results into political action unless there is a wider public perception that there is a substantial environmental problem (Thomas 2004) – for example, as with some of the health and environmental issues associated with lakes noted above.

The institutional arrangements that are development for lake management are important because they set the regulatory context in which certain activities and uses are allowed to occur, as well as often stating the water quality targets both in the lake as well as in the catchment area (e.g. Premazzi et al. 2003). Historical accounts of such institutional arrangements can also provide indications of changing attitudes towards the management of lacustrine systems and their environment, as well as identify the historical roots of contemporary management issues (e.g. Ostendorp et al. 2003; McMahon & Farmer 2004). Although it is readily apparent that land-use in the lake catchment area will affect lake quality, it should be emphasised that the capacity to undertake certain land and water uses are usually subject to regulation and management control.

Introduction to Lacustrine Tourism Systems

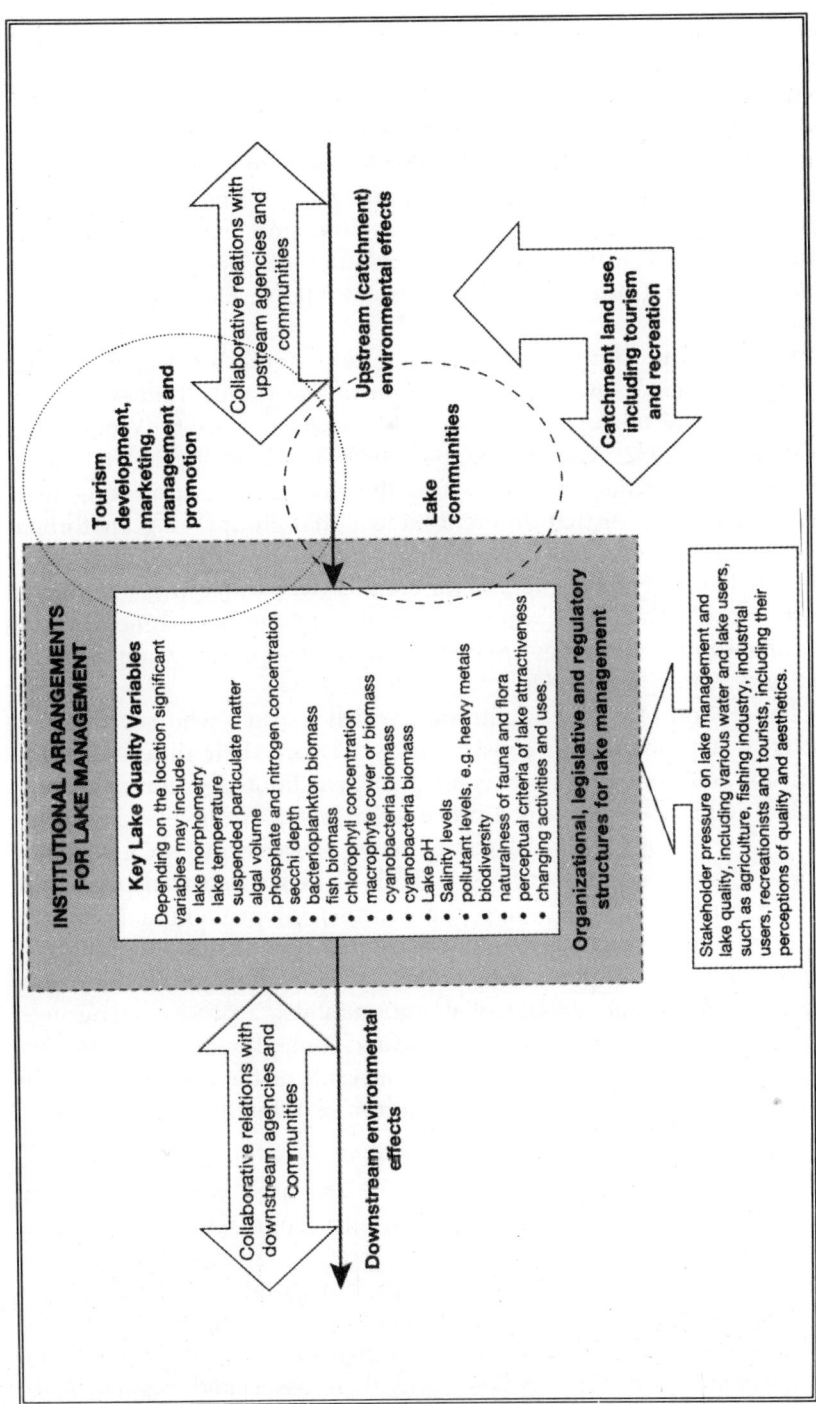

Figure 1.2 Elements in integrated lacustrine management

Therefore, the institutional arrangements, as well as the use and non-use of resources should not be seen as somehow value-free, but instead should be regarded as subject to the values and interests of various stakeholders, including agricultural and industrial groups, such as the fishing and tourism industries, as well as the interests of various community and recreational groups. Where there are many different demands on scarce lake resources, lake management therefore tends to be highly political in terms of the capacities of resource managers to meet stakeholder interests (Cortner & Moote 1999; see Hall & Stoffels, Chapter 11, this volume).

Obviously, the tourism industry is a very significant stakeholder in lake management. However, it is only one of many and, at times, tourism and recreational interests will be at odds with other users (Wood 1982; Wood & Hooy 1982; Warren 1988; Bell 2000; Hall & Shelby 2000). Indeed, at times local communities may feel that tourists are impacting their own quality of use or that different recreational groups – e.g. motorboat users and kayakers – may come into conflict over the same body of water. Increasingly, the potential for conflict over scarce water resources and the need to maintain water quality throughout a catchment means that collaborative strategies are being sought as a part of water management (e.g. Hartig et al. 1998).

Collaboration is a process through which 'parties who see different aspects of a problem can constructively explore their differences and search for solutions that go beyond their own limited vision of what is possible' (Gray 1989: 5). It is a process driven by multiple stakeholders that usually involves several components: (1) agreeing on a common purpose; (2) ensuring the process is both inclusive and transparent; (3) allowing participants to design the process; (4) promoting joint fact-finding and creative problem-solving; (5) insisting on accountability; (6) developing an action plan; and (7) developing collaborative leadership (Margerum & Whitall 2004). Collaboration is also a process being used increasingly to address complex, diffuse and contentious natural resource management and tourism problems as a result of the confluence of the following factors (Michaels 2001; Margerum & Whitall 2004: 408):

- Increased competition for natural resources, including the recognition of ecological uses, has led to increased conflict in natural resource management, with collaborative approaches being seen as an alternative to expensive legal approaches to resolving conflict.
- As understanding of the complexity of natural and social systems increases, more integrated responses to management must be developed that include a wider range of government and non-government decision-makers and their associated capacities for information and analysis.

- Many current environmental problems are a result of diffuse actions such as stormwater run-off, land-use change and habitat modification, which are not well addressed through traditional regulatory actions. Therefore, a collaborative approach that generates broad-scale participation is often better suited to fostering understanding and commitment among a large number of decision-makers.
- There are many different agencies and organisations with overlapping responsibilities and it is argued that a collaborative approach offers an opportunity to reduce waste and duplication, reduce conflict and share data and expertise.

Nevertheless, it should be noted that while collaborative approaches to resource management and tourism have substantial support, there are also significant weaknesses in the approach that suggest that it should be seen as complimentary with regulatory approaches rather than substitute (Cortner & Moote 1999; Wondolleck & Yaffee 2000; McCloskey 2001). Indeed, it may even be argued that regulatory exigencies may serve to encourage collaborative behaviour from stakeholders, such as industry, that would not otherwise collaborate. For example, following their study of ecosystem management in the Lachlan Catchment, New South Wales, Australia, Hillman et al. note that

> whatever social capital gains may occur [as a result of collaborative processes], the resilience and stability needed for the effective operation of an institution at scales required in complex adaptive ecosystem management can only be realistically achieved through legislative underpinning. (2003: 234)

Margerum and Whitall (2004) identify a number of other issues with respect to collaboration and stakeholder attitudes and perceptions (also see Stein et al. 1999; Cantrill et al. 2000) that creates tensions between parties, including:

- the relationship between lay knowledge and expert knowledge;
- the relationship between scale and inclusiveness in collaborative decision-making;
- the relationship between the time frame at which management and stakeholders operate and the time frame in which ecological and scientific data can be provided;
- the significant transaction costs inherent in collaborative management strategies.

These issues were highlighted in Margerum and Whitall's (2004) study of collaborative strategies in environmental management in the Rogue River Basin in Oregon, USA and which also point towards issues in integrated lacustrine management, noting that 'collaboration provides a framework for integrating environmental management activities, but in practice there are substantial barriers and difficulties' (2004: 424). Similarly, a study of collaboration in environmental management in British Columbia (Frame *et al*. 2004: 76), emphasised that collaborative planning is not a panacea in itself, although it is 'a feasible and valuable tool for the resolution of environmental conflicts'. Therefore, while more collaborative approaches are likely to become significant in lake management (e.g. Hartig *et al*. 1998; Erkkila, Chapter 12, this volume) they should not be seen as a substitute for regulatory approaches that help maintain lake and water quality by controlling land-use and activities in the catchment area (Yamamoto & Nakamura 2004). Indeed, it should be emphasised that where regulatory approaches have failed with respect to maintaining lake quality (e.g. Mulqueen 2002; Magadza 2003), it is often a result of poor implementation of regulations, rather than necessarily an inherent problem with regulation itself.

Outline of this Book

This chapter has emphasised that, while tourism and recreation are an important element in lake management, the environmental, social, economical and political dimensions of lacustrine systems necessitate an integrated approach that sees tourism as one, albeit significant component of lake management. This is significant not just for the well-being of lakes and their associated communities, but also for tourism itself as tourism both affects and is affected by lake quality. An understanding of the values associated with lakes and their landscapes is also important because they act as the basis for lakes serving as a destination resource, an issue that is the focus of the following chapter by Cooper that presents an overview of the issues relating to access to lakes as tourism and recreation resources and provides a range of examples of good practice across the world.

Chapters 3–5 detail some of the historical dimensions of lakes and lake tourism and the cultural context within which lakes are perceived as resources. National parks are a familiar expression of the desire to promote tourism through preserving land as part of a natural heritage to be shared with present and future generations of visitors. Lakes are sometimes included in these publicly held reservations, with a few, such as Crater Lake in Oregon, serving as central features. Whether and how they maintain 'iconic' status is the focus of Chapter 3 by Mark, which identifies three stages in cultural perception that affected tourism and park development at Crater Lake. Chapter 4 by Vepsäläinen and Pitkänen details the changing historical dimensions of lake tourism in Savonlinna in Finland and connects the representation of the lake to broader national myths and representations. Chapter 5 by Carr examines the cultural significance of lakes in comparisons of Maori and non-Maori values in association with national parks lakes in New Zealand.

Chapters 6–9 examine profiles of lake and water-based tourist activities and profiles in three European countries: Finland, the Netherlands and Poland. Chapter 6 by Tuohino utilises the notion of spirit of place to describe the potential tourism-marketing value of lakes in the Finnish context. Chapter 7 by Goossen details lake tourism in the Netherlands and outlines some of the changes that have occurred in water-based tourism over time in that country. Chapter 8 by Furgala-Selezniow et al. discusses water tourism issues in the Ostroda–Elblag Canal in Poland and also highlights the connectivity of issues such as heritage, institutional arrangements, water quality and regional development with respect to lake tourism. Chapter 9 by Sievänen and Neuvonen profiles boat recreation in Finland and provides a contrasting macro-perspective to the earlier chapter by Tuohino.

Chapters 10–12 detail planning and management considerations with respect to lake tourism. Chapter 10 by Gartner discussed tourism development issues and trends relevant to lake tourism, along with case studies drawn from the research of Minnesota travellers into and through that state's lake areas. Trends discussed include consumption centre development, public involvement, competition, participation rates and product development. Chapter 11 by Hall and Stoffels details sustainable development and management issues with respect to lake tourism in New Zealand and highlights some of the issues associated between water quality as a result of catchment activities and the value of lakes as a tourism and recreational resource. The chapter also details a recent national strategy with respect to freshwater resources that seeks to provide a more integrated approach to freshwater management. Chapter 12 by Erkilä identifies the need for integrated approaches towards marketing and developing lake destination areas, and highlights the value of creating win–win strategies with respect to satisfying local community needs as well as those of tourists and vacation home-owners. The final chapter, Chapter 13 by Hall and Härkönen, details likely future issues with respect to lake tourism, integrated lake management and environmental change, and the impacts that these may have on lake tourism destinations.

Although lakes have been an under-researched aspect of tourism studies, there is little doubting their significance for tourism in many parts of the world. That significance has grown as people's leisure mobility has increased and the amenity values of lakes has also become significant for lifestyle-related migration including retirement migration and second-home development. In examining the various literatures, it is also apparent that while the subject of lakes has been relatively little researched with respect to tourism in most parts of the world, so it is that lake studies have often failed to study adequately the role of tourism in lacustrine systems. A fundamental aspect of integrated resource management is that there needs to be knowledge transfer not only between stakeholders and managers, but also those, such as researchers, whose specialised knowledge underpins lake management and often serves to set the structures and processes by which management occurs. We therefore hope that this book will be of interest not only to students of tourism, but to all those who have a commitment to maintaining or enhancing the qualities of lakes and lacustrine systems in a world in which water has become an increasingly scarce and sought-after resource.

References

Adeel, Z. (2003) Focus on new water issues – perspectives at the end of the International Year of Freshwater. *Global Environmental Change* 14, 1–4.

Anderson, H.A., Hanrahan, L.P., Smith, A., Draheim, L., Kanarek, M. and Olsen, J. (2004) The role of sport-fish consumption advisories in mercury risk communication: a 1998–1999 12-state survey of women age 18–45. *Environmental Research* 95 (3), 315–24.

Beehler, G.P., Weiner, J.M., McCann, S.E., Vena, J.E. and Sandberg, D.E. (2002) Identification of sport fish consumption patterns in families of recreational anglers through factor analysis. *Environmental Research* 89 (1), 19–28.

Bell, P.J.P. (2000) Contesting rural recreation: the battle over access to Windermere. *Land-use Policy* 17 (4), 295–303.

Brennan, M. (1987) *Wordsworth, Turner and Romantic Landscape*. Columbia, SC: University Press of South Carolina.

Bright, A., Waas, J.R. and Innes, J. (2004) Correlations between human-made structures, boat-pass frequency and the number of New Zealand dabchicks (*Poliocephalus rufopectus*) on the Rotorua Lakes, New Zealand. *New Zealand Journal of Ecology* 28 (1), 137–42.

Bruce, M.G., Curtis, M.B., Payne, M.M., Gautom, R.K., Thompson, E.C., Bennett, A.L. and Kobayashi, J.M. (2003) Lake-associated outbreak of *Escherichia coli* O157:H7 in Clark County, Washington, August 1999. *Archives of Pediatrics & Adolescent Medicine* 157 (10), 1016–21.

Buck, G.M., Mendola, P., Vena, J.E., Sever, L.E., Kostyniak, P., Greizerstein, H., Olson, J. and Stephen, F.D. (1999) Paternal Lake Ontario fish consumption and risk of conception delay, New York State Angler Cohort. *Environmental Research, Section A*, 80 (2), S13–S18.

Buck, G.M., Vena, J.E., Schisterman, E.F., Dmochowski, J., Mendola, P., Sever, L.E., Fitzgerald, E., Kostyniak, P., Greizerstein, H. and Olson, J. (2000) Parental consumption of contaminated sport fish from Lake Ontario and predicted fecundability. *Epidemiology* 11 (4), 388–93.

Buzard, J. (1983) *The Beaten Track: European Tourism, Literature and the Ways to 'Culture', 1800–1918*. Oxford: Clarendon Press.

Cantrill, J., Potter, T. and Stephenson, W. (2000) Protected areas and regional sustainability: Surveying decision makers in the Lake Superior Basin. *Natural Resources Journal* 40 (1), 19–45.

Carabelli, F.A. (2001) A proposal for the development of tourism in the forested landscapes of Tierra del Fuego, Patagonia, Argentina. *Tourism Analysis* 6 (3/4), 185–202.

Castle, K. (2003) Out of the blue. *Ski Area Management* 42 (4), 44–5, 52–3.

Chamot, E., Toscani, L. and Rougemont, A. (1998) Public health importance and risk factors for cercarial dermatitis associated with swimming in Lake Leman at Geneva, Switzerland. *Epidemiology and Infection* 120 (3), 305–14.

Chape, S., Blyth, S., Fish, L., Fox, P. and Spalding, M. (compilers) (2003) *2003 United Nations List of Protected Areas*. IUCN, Gland, Switzerland and Cambridge, UK and United Nations Environment Programme–World Conservation Monitoring Centre, Cambridge, UK.

Christensen, N.A. and Cole, D.N. (2000) Leave no trace practices: Behaviours and preferences of wilderness visitors regarding use of cookstoves and camping away from lakes. In D.N. Cole, S.E. McCool, W.T. Borrie and J. O'Loughlin (eds) *Proceedings – Rocky Mountain Research Station, USDA Forest Service*, No. RMRS-P-15 (4), 77–85). Missoula, MT: Aldo Leopold Wilderness Research Institute, US Department of Agriculture, Forest Service, Rocky Mountain Research Station.

Cole, D.C., Kearney, J., Sanin, L.H., Leblanc, A. and Weber, J.P. (2004) Blood mercury levels among Ontario anglers and sport-fish eaters. *Environmental Research* 95 (3), 305–14.

Connelly, N.A., Brown, T.L., Knuth, B.A. and Wedge, L. (1999) Changes in the utilisation of New York's great lakes recreational fisheries. *Journal of Great Lakes Research* 25 (2), 347–54.

Cortner, H.J. and Moote, M.A. (1999) *The Politics of Ecosystem Management*. Washington, DC: Island Press.

Courval, J.M., DeHoog, J.V., Stein, A.D., Tay, E.M., He J. Humphrey, H.E.B. and Paneth, N. (1999) Sport-caught fish consumption and conception delay in licenced Michigan anglers. *Environmental Research, Section A* 80 (2), S183–S188.

Curtis, J.A. (2003) Demand for water-based leisure activity. *Journal of Environmental Planning and Management* 46 (1), 65–77.

Dellinger, J.A. (2004) Exposure assessment and initial intervention regarding fish consumption of tribal members of the Upper Great Lakes Region in the United States. *Environmental Research* 95 (3), 325–40.

Department of the Environment (2003) *Western Australia's Waterways: Understand, Protect, Restore* [poster]. Perth: Department of the Environment.

Diniyati, D. (2002) Segmentasi pasar wisata ekologi di sekitar danau Toba (Market segment of ecotourism surrounding the Toba lake). *Buletin Penelitian Kehutanan – Pematang Siantar* 18 (1), 48–57.

Dredge, D. (2001) Leisure lifestyles and tourism: Socio-cultural, economic and spatial change in Lake Macquarie. *Tourism Geographies* 3 (3), 279–99.

Edington, J.M. and Edington, M.A. (1986) *Ecology, Recreation and Tourism*. Cambridge: Cambridge University Press.

Egan, K., Herriges, J., Kling, C. and Downing, J. (2004) Valuing water quality in midwestern lake ecosystems. *IOWA Ag Review* 10 (3), 4–5.

Eiswerth, M.E., Englin, J., Fadali, E. and Shaw, W.D. (2000) The value of water levels in water-based recreation: A pooled revealed preference/contingent behaviour model. *Water Resources Research* 36 (4), 1079–86.

Fadali, E. and Shaw, W.D. (1998) Can recreation values for a lake constitute a market for banked agricultural water? *Contemporary Economic Policy* 16 (4), 433–41.

Falk, C , Hanrahan, L. anderson, H.A., Kanarek, M.S., Draheim, L., Needham, L. and Patterson, D., Jr (1999) Body burden levels of dioxin, furans and PCBs among frequent consumers of Great Lakes sport fish. *Environmental Research, Section A* 80 (2), S19–S25.

Feldman, K.A., Mohle-Boetani, J.C., Ward, J., Furst, K., Abbott, S.L., Ferrero, D.V., Olsen, A. and Werner, S.B. (2002) A cluster of *Escherichia coli* O157:

Nonmotile infections associated with recreational exposure to lake water. *Public Health Reports* 117 (4), 380–5.

Frame, T.M., Gunton, T. and Day, J.C. (2004) The role of collaboration in environmental management: An evaluation of land and resource planning in British Columbia. *Journal of Environmental Planning and Management* 47 (1), 59–82.

Gray, B. (1989) *Collaborating: Finding Common Ground for Multiparty Problems.* San Francisco, CA: Jossey-Bass.

Guo, H.C., Liu, L., Huang, G.H., Fuller, G.A., Zou, R. and Yin, Y.Y. (2001) A system dynamics approach for regional environmental planning and management: A study for the Lake Erhai Basin. *Journal of Environmental Management* 61 (1), 93–111.

Hadwen, W.L. and Arthington, A.H. (2003) The significance and management implications of perched dune lakes as swimming and recreation sites on Fraser Island, Australia. *Journal of Tourism Studies* 14 (2), 35–44.

Hadwen, W.L., Arthington, A.H. and Mosisch, T.D. (2003) The impact of tourism on dune lakes on Fraser Island, Australia. *Lakes and Reservoirs: Research and Management* 8 (1), 15–26.

Haigh, M.J., Lansky, L. and Hellin, J. (2004) Headwater deforestation: A challenge for environmental management. *Global Environmental Change* 14, 51–61.

Hall, T. and Shelby, B. (2000) Temporal and spatial displacement: Evidence from a high-use reservoir and alternate sites. *Journal of Leisure Research* 32 (4), 435–56.

Härkönen, T. (ed.) (2003) *International Lake Tourism Conference, 2–5 July, 2003, Savonlinna, Finland.* Savonlinna: Savonlinna Institute for Regional Development and Research, University of Joensuu.

Hartig, J.H., Zarull, M.A., Heidtke, T.M. and Shah, H. (1998) Implementing ecosystem-based management: Lessons from the Great Lakes. *Journal of Environmental Planning and Management* 41 (1), 45–75.

He, J., Stein, A.D., Humphrey, H.E.B., Paneth, N., Courval, J.M. (2001) Time trends in sport-caught great lakes fish consumption and serum polychlorinated biphenyl levels among Michigan anglers, 1973–1993. *Environmental Science & Technology* 35 (3), 435–40.

Heathcote, I.W. (1998) *Integrated Watershed Management: Principles and Practice.* Chichester: John Wiley & Sons.

Henderson, J.E., Kirk, J.P., Lamprecht, S.D. and Hayes, W.E. (2003) Economic impacts of aquatic vegetation to angling in two South Carolina reservoirs. *Journal of Aquatic Plant Management* 41, 53–6.

Hickley, P., Arlinghaus, R., Tyner, R., Aprahamian, M., Parry, K. and Carter, M. (2004) Rehabilitation of urban lake fisheries for angling by managing habitat: General overview and case studies from England and Wales. *International Journal of Ecohydrology & Hydrobiology* 4 (4), 365–78.

Hillmann, M., Aplin, G. and Brierley, G. (2003) The importance of process in ecosystem management: Lessons from the Lachlan Catchment, New South Wales, Australia. *Journal of Environmental Planning and Management* 46 (2), 219–37.

Jakus, P.M., Dadakas, D. and Fly, J.M. (1998) Fish consumption advisories: Incorporating angler-specific knowledge, habits and catch rates in a site choice model. *American Journal of Agricultural Economics* 80 (5), 1019–24.

Jin, G.A., Englande, A.J., Jr, Bradford, H. and Jeng, H.W. (2004) Comparison of E. coli, Enterococci and fecal coliform as indicators for brackish water quality assessment. *Water Environment Research* 76 (3), 245–55.

Jin, G., Englande, A.J., Jr and Liu, A. (2003) A preliminary study on coastal water quality monitoring and modeling. *Journal of Environmental Science and Health. Part A, Toxic/Hazardous Substances & Environmental Engineering* 38 (3), 493–509.

Jones, B., Scott, D. and Gössling, S. (2005) Lakes and streams. In S. Gössling and C.M. Hall (eds) *Tourism and Global Environmental Change* (pp. 76–94). London: Routledge.

Karakoc, G., Erkoc, F.Ü. and Katircioglu, H. (2003) Water quality and impacts of pollution sources for Eymir and Mogan Lakes (Turkey). *Environment International* 29 (1), 21–7.

Kassilly, F.N. (2003) Towards promotion of local tourism: The case of Lake Nakuru National Park in Kenya. *African Journal of Ecology* 41 (2), 187–9.

King, J. G. and Arnett, C.M. (1974) Effects of recreation on water quality. *Journal of the Water Pollution Control Federation* 46 (11), 2453–9.

Kline, M.B. (1970) *Beyond the Land Itself: Views of Nature in Canada and the United States*, Cambridge, MA: Harvard University Press.

Kuentzel, W.F. and Heberlein, T.A. (2003) More visitors, less crowding: Change and stability of norms over time at the Apostle Islands. *Journal of Leisure Research* 35 (4), 349–71.

Lake District National Park Authority (1999) *Lake District National Park Management Plan*. Kendal: Lake District National Park Authority.

Lee, H. (2001) Determinants of recreational boater expenditures on trips. *Tourism Management* 22 (6), 659–67.

Lee, H. (2002) Regional flows of recreational boater expenditures on trips in Michigan. *Journal of Travel Research* 41 (1), 77–84.

Lee, H. (2003a) Modeling boaters' choices among boating destinations in Michigan's Great Lakes. *Tourism Analysis* 7 (3/4), 217–28.

Lee, H. (2003b) Estimating recreational boater expenditures on trips and boating use in a wave survey. *Leisure Sciences* 25 (4), 381–97.

Lee, S., Kim, B. and Oh, H. (2002) Evaluation of lake modification alternatives for Lake Sihwa, Korea. *Environmental Management* 29 (1), 57–66.

Leslie, D. (2002) National parks and the tourism sector. *Countryside Recreation* 10 (3/4), 5–10.

Levy, D.A., Bens, M.S., Craun, G.F., Calderon, R.L. and Herwaldt, B.L. (1998) Surveillance for waterborne-disease outbreaks – United States, 1995–1996. *Morbidity and Mortality Weekly Report* 47 (SS-5), 1–33.

Liddle, M.J. and Scorgie, R.A. (1980) The effects of recreation on freshwater plants and animals: A review. *Biological Conservation* 17 (2), 183–206.

Lockwood, R.N., Peck, J. and Oelfke, J. (2001) Survey of angling in Lake Superior waters at Isle Royale National Park, 1998. *North American Journal of Fisheries Management* 21 (3), 471–81.

McCloskey, M. (2001) Is this the course you want to be on? *Society and Natural Resources* 14 (4), 627–34.

McElroy, J.A., Kanarek, M.S., Trentham-Dietz, A., Robert, S.A., Hampton, J.M., Newcomb, P.A., Anderson, H.A. and Remington, P.L. (2004) Potential exposure to PCBs, DDT and PBDEs from sport-caught fish consumption in relation to breast cancer risk in Wisconsin. *Environmental Health Perspectives* 112 (2), 156–62.

McLellan, S.L. and Salmore, A.K. (2003) Evidence for localized bacterial loading as the cause of chronic beach closings in a freshwater marina. *Water Research* 37 (11), 2700–8.

McMahon, G.F. and Farmer, M.C. (2004) Reallocation of federal multipurpose reservoirs: Principles, policy and practice. *Journal of Water Resources Planning and Management* 130 (3), 187–97.

Madrzejowska, K. and Borowski, J. (2003) Sailing tourism and the condition of flora at the banks of Beldany lake. *Annals of Warsaw Agricultural University, Horticulture (Landscape Architecture)* 24, 91–104.

Magadza, C.H.D. (2003) Lake Chivero: A management case study. *Lakes & Reservoirs: Research and Management* 8, 69–81.

Maharana, I., Rai, S.C. and Sharma, E. (2000) Valuing ecotourism in a sacred lake of the Sikkim Himalaya, India. *Environmental Conservation* 27 (3), 269–77.

Mangun, J.C. and O'Leary, J.T. (2001) Macrosociological inquiry and sport fishing. *Society & Natural Resources* 14 (5), 385–97.

Marcouiller, D.W., Kim K. and Deller, S.C. (2004) Natural amenities, tourism and income distribution. *Annals of Tourism Research* 31 (4), 1031–50.

Margerum, R.D. (1999) Integrated environmental management: The elements critical to success. *Environmental Management* 65 (2), 151–66.

Margerum, R.D. and Whitall, D. (2004) The challenges and implications of collaborative management on a river basin scale. *Journal of Environmental Planning and Management* 47 (3), 407–27.

Michaels, S. (2001) Making collaborative watershed management work: The confluence of state and regional activities. *Environmental Management* 27 (1), 27–35.

Mitchell, B. (ed.) (1990) *Integrated Water Management*. London: Belhaven.

Morgan, J., Bornstein, S.L., Karpati, A.M., Bruce, M., Bolin, C.A., Austin, C.C., Woods, C.W., Lingappa, J., Langkop, C., Davis, B., Graham, D.R., Proctor, M., Ashford, D.A., Bajani, M., Bragg, S.L., Shutt, K., Perkins, B.A. and Tappero, J.W. (2002) Outbreak of leptospirosis among triathlon participants and community residents in Springfield, Illinois, 1998. *Clinical Infectious Diseases* 34 (12), 1593–9.

Mosisch, T.D. and Arthington, A.H. (2004) Impacts of recreational power-boating on freshwater ecosystems. In R. Buckley (ed.) *Environmental Impacts of Ecotourism* (pp. 125–54). Wallingford: CABI Publishing.

Mulqueen, J. (2002) The trophic status of Lough Conn: A review. *Farm & Food* 12 (2), 28–30.

Murray, C., Sohngen, B. and Pendleton, L. (2001) Valuing water quality advisories and beach amenities in the Great Lakes. *Water Resources Research* 37 (10), 2583–90.

Newsome, D. Moore, S.S.A. and Dowling, R.K. (2002) *Natural Area Tourism: Ecology, Impacts and Management*. Clevedon: Channel View Publications.

Odada, E.O. (2004) Basic research on the world's lakes lags far behind similar research on the oceans. *Lakes & Reservoirs: Research and Management* 9, 161–2.

Ostendorp, W., Walz, N. and Brüggemann, R. (2003) Problemfeld Seeufer am Beispiel Bodensee: Umsetzung der Uferschutz-Bestimmungen (Conflicts in lake shore protection – example Lake Constance). *Umweltwissenschaften und Schadstoff-Forschung* 15 (3), 187–98.

Ousby, I. (1990) *The Englishman's England: Taste, Travel and the Rise of Tourism*. Cambridge: Cambridge University Press.

Paunio, M., Pebody, R., Keskimäki, M., Kokki, M., Ruutu, P., Oinonen, S., Vuotari, V., Siitonen, A., Lahti, E. and Leinikki, P. (1999) Swimming-associated outbreak of *Escherichia coli* O157:H7. *Epidemiology and Infection* 122 (1), 1–5.

Payne, R.J., Johnston, M.E. and Twynam, G.D. (2001) Tourism, sustainability and the social milieus in Lake Superior's north shore and islands. In S.F. McCool and R.N. Moisey (eds) *Tourism, Recreation and Sustainability: Linking Culture and the Environment* (pp. 315–42). Wallingford: CABI Publishing.

Premazzi, G., Dalmiglio, A., Cardoso, A.C. and Chiaudani, G. (2003) Lake management in Italy: The implications of the Water Framework Directive. *Lakes & Reservoirs: Research and Management* 8, 41–59.

Pickles, K. (2003) The re-creation of Bottle Lake: From site of discard to environmental playground? *Environment and History* 9 (4), 419–34.

Provencher, B., Baerenklau, K.A. and Bishop, R.C. (2002) Finite mixture logit model of recreational angling with serially correlated random utility. *American Journal of Agricultural Economics* 84 (4), 1066–75.

Puczkó, L. and Rátz, T. (2000) Tourist and resident perceptions of the physical impacts of tourism at Lake Balaton, Hungary: Issues for sustainable tourism management. *Journal of Sustainable Tourism* 8 (6), 458–78.

Rabe, B.G. (1996) An empirical examination of innovations in integrated environmental management: The case of the Great Lakes Basin. *Public Administration Review* 56 (4), 372–81.

Reed, J.R. and Parsons, B.G. (1999) Angler opinions of bluegill management and related hypothetical effects on bluegill fisheries in four Minnesota lakes. *North American Journal of Fisheries Management* 19 (2), 515–19.

Reed-Andersen, T., Bennett, E.M., Jorgensen, B.S., Lauster, G., Lewis, D.B., Nowacek, D., Riera, J.L., Sanderson, B.L. and Stedman, R. (2000) Distribution of recreational boating across lakes: Do landscape variables affect recreational use? *Freshwater Biology* 43 (3), 439–48.

Renberg, I., Bindler, R., Bradshaw, E., Emteryd, O. and McGowan, S. (2001) Sediment evidence of early eutrophication and heavy metal pollution of Lake Mälaren, Central Sweden. *Ambio* 30 (8), 496–502.

Stauber, K. (2004) Cost-effective use of visitor-use exclosures for the rejuvination of *Ammophila breviligulata* on Minnesota Point, Minnesota. *Natural Areas Journal* 24 (1), 32–5.

Stein, T.V. Anderson, D.H. and Thompson, D. (1999) Identifying and managing for community benefits in Minnesota state parks. *Journal of Park and Recreation Administration* 17 (4), 1–19.

Steiner, W. and Parz-Gollner, R. (2003) Actual numbers and effects of recreational disturbance on the distribution and behaviour of Greylag Geese (*Anser anser*) in the Neusiedler See – Seewinkel National Park Area. *Journal for Nature Conservation* 11 (4), 324–30.

Szylvian, K.M. (2004) Transforming Lake Michigan into the 'world's greatest fishing hole': The environmental politics of Michigan's Great Lakes sport fishing, 1965–1985. *Environmental History* 9 (1), 102–27.

Tambuk-Giljanovic, N. (2003) The water quality of the Vrgorska Matica River. *Environmental Monitoring and Assessment* 83 (3), 229–53.

Thomas, R.L. (2004) Management of freshwater systems: The interactive roles of science, politics and management and the public. *Lakes & Reservoirs: Research and Management* 9, 65–73.

Truly, D. (2002) International retirement migration and tourism along the Lake Chapala Riviera: Developing a matrix of retirement migration behaviour. *Tourism Geographies* 4 (3), 261–81.

Tuohino, A. and Pitkänen, K. (2004) The transformation of a neutral lake landscape into a meaningful experience – interpreting tourist photos. *Journal of Tourism and Cultural Change* 2 (2), 77–93.

Udvardy, M.D.F. (1975) *A Classification of the Biogeographical Provinces of the World*, IUCN, Occasional Paper 18. Gland, Switzerland.

Vogelsong, H. and Ellis, C. (2001) *Assessing the Economic Impact of Ecotourism Developments on the Albermarle/Pamlico Region*. Washington, DC: Economic Development Administration, United States Department of Commerce.

Vreugdenhil, D., Terborgh, J., Cleef, A.M., Sinitsyn, M., Boere, G.D., Archaga, V.L. and Prins, H.H.T. (2003) *Comprehensive Protected Areas System Composition and Monitoring*, World Institute for Conservation and Environment, Shepherdstown, WV.

Walton, J.K. and McGloin, P.R. (1981) The tourist trade in Victorian lakeland. *Northern History* 17, 153–82.

Warren, R. (1988) Lake maintenance brings quality recreation. *Parks and Recreation* 23 (4), 38–40.

Wondolleck, J.M. and Yaffee, S.L. (2000) *Making Collaboration Work: Lessons From Innovations in Natural Resource Management*. Washington, DC: Island Press.

Wood, J. (1982) Lake carrying capacity. *Recreation Australia* 2 (1), 30–40.

Wood, J. and Hooy, T. (1982) Planning for quality sailing experiences on Canberra's lakes. *Australian Parks and Recreation* 18 (5), 14–21.

Wynne, D., Shteinman, B., Hochman, A. and Ben-Dan, T.B. (2004) The spatial distribution of enteric bacteria in the Jordan River-Lake Kinneret contact zone. *Journal of Toxicology and Environmental Health. Part A* 67 (20/22), 1705–15.

Yamamoto, K. and Nakamura, M. (2004) An examination of land use controls in the Lake Briva watershed from the perspective of environmental conservation and management. *Lakes & Reservoirs: Research and Management* 9, 217–28.

Yan, S., Zhang C., Xu, Y. and Kan, Y. (2000) Some concepts for developing the tourist resources in Yili Kazak Autonomous Prefecture, Xinjiang, China. *Arid Land Geography* 23 (1), 1–6.

Yang, T., Yan, S., Yang, Z. and Xu, Y. (1997) Preliminary study on the sustainable development of tourist industry in Tianchi Lake scenic spot, Tianshan Mountains. *Arid Land Geography* 20 (4), 47–53.

Yoder, J.S., Blackburn, B.G., Craun, G.F., Hill, V., Levy, D.A., Chen, N., Lee, S.H., Calderon, R.L. and Beach, M.J. (2004) Surveillance for waterborne-disease outbreaks associated with recreational water – United States, 2001–2002. *Morbidity and Mortality Weekly Report* 53 (SS-8), 1–21.

Zhang C. (1999) Division and development strategy of tourism in Altay prefecture, Xinjiang. *Arid Land Geography* 22 (1), 20–6.

Zyl, H., van Store, T. and Leiman, A. (2000) The recreational value of viewing wildlife in Kenya. In J. Rietbergen-McCracken and H. Abaza (eds) *Environmental Valuation: A Worldwide Compendium of Case Studies* (pp. 35–52). London: Earthscan Publications Ltd.

Zyll de Jong, M.C., van Lester, N.P., Korver, R.M., Norris, W., Wicks, B.L. (2000) Managing the exploitation of brook trout, *Salvelinus fontinalis* (Mitchill), populations in Newfoundland lakes. In I.G. Cowx (ed.) *Management and Ecology of Lake and Reservoir Fisheries. Proceedings of the Symposium and Workshop on Management and Ecology of Lake and Reservoir Fisheries, Hull, UK, April 2000*, (pp. 267–83). International Fisheries Institute, University of Hull in cooperation with the European Inland Fisheries Advisory Commission, Hull.

Chapter 2
Lakes as Tourism Destination Resources

CHRIS COOPER

Introduction

Lakes are open water bodies, ponds, dams or reservoirs on the surface of the earth, representing a valuable resource utilised for a variety of human activities. Globally, lakes have an uneven distribution, dominantly found in upland regions, with Canada alone claiming around half of all the world's lakes. The water quality of lakes is critical to their attractiveness as recreation and tourism resources, with many lakes in the world being salt water rather than fresh water (such as the Caspian and the Dead Seas). The landscape setting of lakes is also an important aspect of their attractiveness, coming in a variety of forms:

- glacial lakes;
- caldera lakes;
- underground lakes;
- rift valley lakes;
- ox bow lakes;
- artificial lakes and reservoirs.

As can be seen from this list, lakes are significant landscape features; the largest of the earth's lakes – Lake Baikal, for example, or Lake Victoria – can be seen from space (Jorgensen & Matsui 1997). Lakes can also be artificially created landscape features and here, landscape aesthetics is an increasing element of planning and budget. Nonetheless, despite their scale and significance, lakes are relatively transient landscape features, eventually disappearing through climate change, eutrophication or erosion – their life can be measured in terms of thousands of years rather than millions.

As well as natural features, lakes are also critical in supporting human life and have a social significance dating back thousands of years; indeed, lakeshores are significant archaeological resources, having supported settlements for centuries. In terms of supporting life, lakes are a vital resource, representing water catchment basins for almost 40% of the world's land, as well as acting as focal points for tourism and recreation. Yet, while lakes have traditionally been used to support human life, they are increasingly being seen as areas for recreation and tourism, attracting visitation through their landscape features, flora and fauna, and cultural attractions.

In terms of the analysis of tourism destinations, lakes are significant. They represent the attraction element of a destination as they generate the visit to the region. As such, they are often the core of a destination's attractiveness – as at Lake Balaton in Hungary or the English Lake District. Lakes represent resource-based tourism attractions, demanding high levels of management and coordination between users. Yet it is the very nature of tourism destinations that poses the greatest threat to lakes as tourism resources. Not only is demand often focused seasonally and at weekends, but also the fragility of lake ecosystems exacerbates their vulnerability to, often unintentional, tourism impacts.

In considering lakes as tourism destinations, the World Tourism Organisation's (WTO) (2002, n.p.) attempts to draft a definition of 'tourism destination' are helpful:

> A local tourism destination is a physical space in which a visitor spends at least one overnight. It includes tourism products such as support services and attractions and tourism resources within one day's return travel time. It has physical and administrative boundaries defining its management and images and perceptions defining its market competitiveness. Local tourism destinations incorporate various stakeholders often including a local community and can nest and network to form larger destinations.

This initial approach to analysing destinations provides an insight into the management issues of lakes as tourism destinations (Cooper *et al.* 1998). Taking the lead of the WTO, we can think of lake destinations as being:

- cultural appraisals, comprised of images and perceptions;
- perishable and vulnerable to change;
- used by multiple users or stakeholders;
- complex amalgams in need of management.

The remaining sections of this chapter are structured around the four points above.

Lakes as Cultural Appraisals

There is no doubt that society places a high value on lakes (Fadali & Shaw 1998). All societies depend on the availability of fresh water for their survival and development, demanding that the use and management of lakes is sustainable. Yet the increasingly complex nature of the many activities undertaken at lakes, in addition to the supply of fresh water, implies a multiplicity of stakeholders. Each group of stakeholders will have a different level of access and control over scarce lake resources – water, land or recreational space – and this inequality of access causes a form of stakeholder stratification. With recreation and tourism often the most recent use of lakes, their priorities are often low down in this stratification.

Recreation and tourism resources are effectively cultural appraisals – society has to view them as worthwhile and attractive places to visit. Of course, tastes and preferences change and so the value that we place upon lakes as tourism resources has altered over time. For example, changes in landscape tastes have placed severe pressures on the lakes and mountain scenery of, say, alpine Italy, Austria and Switzerland. These lakes are a classic example of the cultural appraisal of tourism resources, having been 'discovered' by the Romantic writers and composers of the early 19th century. The Italian lakes combine both natural and cultural tourism resources, with their stunning landscape settings complemented by castles, villas and gardens. Yet in the 21st century their accessibility adds to their vulnerability as tourism destinations, as does the acute seasonal peaking of demand in July and August. Ironically, though, it is the over-development of winter sports resorts in the mountains that is impacting more severely upon the lakes with increased mudslides and floods (Boniface & Cooper 2001).

The local community is a key consideration of the cultural appraisal of lakes as recreation and tourism resources. They are not only part of the attraction of many lake destinations, but they also are able to benefit from the commercial opportunities offered by tourism. Here, they may be dependent on other elements of destination features such as access. For example, the 1500 inhabitants of Taquile Island on Lake Titicaca were isolated until ready access by motorboat from the nearest Peruvian port of Puno, opened up the island to visitors. The Taquileans have since developed a tourism industry based on home-stays, a unique rotating ownership system of local restaurants and the equitable division of profits made from shops and restaurants (Collins 1997).

Despite the critical need to consider the cultural dimension, management of lakes has generally been on the basis of biophysical and economic aspects, rather than from a social or cultural point of view. Klessig (2001) argues that citizens should be consulted on the role of lakes. As society has increasingly recognised the value of lakes, so too has the level of interest in local communities becoming involved in the management and protection of lakes. Citizen participation in lake management and custodianship is growing, particularly in the USA and Canada. Involvement of citizens in the management of lakes is seen in the Australian context through integrated catchment management, in Wisconsin with the Wisconsin Lakes Partnership and in the Great Lakes with the Great Lakes United coalition of citizen groups. The advantages of community involvement are clear. Natural resource experts educate citizens as to sound management practices, while the role of citizen involvement broadens the context and scope of lake management. Of course, some citizen groups have strong vested interests, but public agencies can balance these groups – for example, in Wisconsin, taxes paid on fuel by lake motorboat users is used to fund lake management activities rather than roads (Klessig 2001). Increasingly, such citizen groups can access government support through funding and information access. Two groups (mentioned above) are prominent here:

(1) *The Wisconsin Lakes Partnership* A collaborative group of organisations dedicated to the protection of Wisconsin's lakes for future generations. The Partnership has developed a model of lake classification and watershed restoration to achieve lake protection goals.
(2) *Great Lakes United* An international coalition dedicated to preserving and restoring the Great Lakes/St Lawrence River ecosystem. Great Lakes United is made up of organisations of environmentalists, conservationists, hunters and anglers, labour unions, community groups and citizens of the United States, Canada and First Nations and Tribes. The organisation was founded in 1982 and has developed a range of policy initiatives, education programmes and citizen's action groups to assure:

- clean water and clean air for all citizens;
- better safeguards to protect the health of people and wildlife; and
- a conservation ethic that will leave healthy Great Lakes.

The Vulnerability of Lake Destinations

All tourism destinations are vulnerable to change by forces both external to tourism and also by tourism itself. Lakes as tourism destinations are particularly vulnerable to change by recreation and tourism use as they represent fragile eco-systems. However, external threats endanger the integrity of lakes and thus their attraction as recreation and tourism destinations. From the point of view of these external threats, the United Nations Environment Programme has identified six key environmental problems facing lakes (Jorgensen & Matsui 1997):

(1) Lowering of water levels due to over-use of water (e.g. the Aral Sea, Kazakhstan and Uzbekistan).

(2) Rapid siltation due to increased run-off (e.g. Lake Dongting Hu, China).
(3) Acidification of water caused by acid precipitation (Lake Biwa, Japan).
(4) Contamination of water by toxic pollutants.
(5) Eutrophication due to inflow of nutrients.
(6) In some cases, collapse of aquatic ecosystems.

Once these processes reach an advanced stage they are irreversible, meaning that they must be identified and managed at an early stage. It also has to be remembered that, unlike the sea with its tidal scour, lakes have no natural cleansing mechanism and so are more susceptible to water pollution caused by tourism and other uses (Boniface & Cooper 2001). Turning to the specific effects of tourism and recreation, impacts vary according to two sets of factors:

(1) The physical features of the lake – location, type and size of the lake.
(2) Tourism variables – here, Hadwen et al. (2003) have developed a tourist pressure index (TPI) for dune lakes on Fraser Island, Australia. This clearly identified vulnerable lakes on the basis of accessibility, facilities and prominence in advertising campaigns.

Water Pollution

The entry of waste from recreational developments and activities into lakes is a major cause of environmental quality problems (Newsome et al. 2002). The impact is greater on smaller enclosed lakes with limited exchanges of water. Nutrients and suspended solids enter the water, decreasing the oxygen content through the growth of algae feeding on the nutrients. In turn, this can lead to the death of fish. In other words, tourists adversely affect the ecology of lakes through the addition of nutrients and other chemicals to the water (Hadwen et al. 2003). King and Arnett (1974) found that recreation significantly affects water quality, although the severity of the impact depends upon the local geology and soils. Parkes (1974) found significant reductions of demand for water-based recreation in lakes where algae are present – and this raises the issue of the multiple use of lakes for recreation, water supply and other users. The key question here is: which of the user groups should determine water quality and should this be assessed by scientific or social attitude measures?

Shoreline Impacts

In addition to impacts on the lake water itself, Jorgensen and Loffler (1990) note that the shoreline or littoral zone is the zone of highest recreation and tourism impact. This zone is vulnerable, particularly in shallow lakes, not only from an economic point of view, but also in terms of landscape quality and fragile resources – for example, shorelines are often the site of archaeological remains having been areas of settlement, as with Neolithic and Bronze Age farmers in Europe.

To summarise, it is possible to identify a schedule of the impacts of recreation and tourism on bodies of water (King & Arnett 1974; Liddle & Scorgie 1980; Edington & Edington 1986; Newsome et al. 2002):

- damage to fauna by boat propellers;
- recreational fishing and the introduction of exotic species (e.g. rainbow trout in Lake Titicaca);
- lead poisoning of water birds;
- algal growths;
- discharge of sewage and garbage;
- outboard motor fuel and oil;
- camping and other activities accelerating soil erosion and run-off;
- disturbance to fish reproduction and egg hatching;
- bank erosion and loss of plants due to wave action and boat contact;
- reduction of species diversity;
- trampling on lakeshores;
- increased turbidity due to boating, water skiing, jet skis and propeller action;
- impact on water birds through noise and disturbance.

Finally, it is important that the levels of impacts are monitored and this raises issues of data collection as many impacts on lakes are not as localised as impacts upon sites, footpaths or picnic areas – upstream and downstream considerations must be taken into account in the estimation of impacts (Newsome et al. 2002).

Multiple Use of Lake Destinations

Recreation and tourism are often the most recent uses for lakes and reservoirs and have to be 'fitted in' with other uses. Conflicts arise here with other stakeholders and the relatively recent practice of including the local community in the stakeholder groups has complicated management agendas at popular lake destinations. In addition, the fact that tourism and recreation are often seasonal in their use of lakes and reservoirs also reduces their bargaining power. Lake Balaton in Hungary, for example, is a highly seasonal lake destination as it is covered by ice in winter.

Tourism and recreation are commonly secondary activities at lakes and reservoirs where the primary use is water supply, flood mitigation or power supply (Pitts & Anderson 1985). They are therefore subject to both stringent management controls and activity restrictions to ensure that tourism and recreation is compatible with the primary use of the water body. As this principle has become accepted, recreational use is planned from the start of the design of reservoirs. This involves multi-agency cooperation and the consideration of recreation on 'equal' terms with other uses.

The recreational use of artificial lakes and reservoirs is more recent than for natural lakes. In England and Wales, for example, despite artificial water schemes dating back to the early 1600s, the use of reservoirs for recreation and public access only dates to the later 19th century (WSAC 1977). Here, the classic conflict over recreation access versus fears for contamination of the catchment areas for the water supply of nearby cities led to controversy and court cases. It was not until the early years of the 20th century that the water surface itself began to be used for recreation – mainly fishing, although widespread use of the water surface for other recreation was discouraged.

This conflict between different lake uses is problematic and demands careful multiple use planning and conflict resolution. Examples of other uses of lakes are:

- water supply;
- fishing;
- flood control;
- irrigation;
- nature conservation;
- transportation;
- power generation.

Even in terms of recreation and tourism, many lake uses conflict with each other – public swimming conflicts with water supply, public access may conflict with private land ownership, motorised boating conflicts with swimming, canoeing with motorboats – and indeed both recreation and tourism can conflict with wildlife and landscape conservation. From a recreation and tourism point of view, conflicts between users reduce satisfaction (Wood & Hooy 1982). Conflict arises from a variety of sources, such as the inconsiderate behaviour on and off the lake, the poor management of car parking and boat launching, or the fact that recreation and tourism users have different needs – for example, recreational sailors need 50% more water area than competitive sailors, while learner sailors need 100% as much (Wood & Hooy 1982).

Wang and Dawson (2000) examined the conflict between recreation users on the Great Lakes. They defined recreation conflict as 'interference to a user, who is trying to achieve a goal in a recreation activity and the interference is due to another recreation user's behaviour' (2000: 1).

Conflict can be linked to the user's goals in behaviour, their activity style, tolerance of other users and sensitivity to conflict. For example,

Jaakson (1989) suggests that recreation users of lakes can be divided into *sensitive* (adversely affected by others) and *tolerant* (not adversely affected by others, but may adversely affect others). Wang and Dawson found that 'asymmetric conflicts' are common in relation to water-based recreation – in other words, only one party is impacted upon. This suggests that management solutions must identify the various users and resolve conflicts for individual groups – failure to resolve these conflicts leads to exacerbated situations. Solutions include the education of users, space and distance guidelines, separating craft and users (including landowners), the relocation of the land-based element of the offending activity (e.g. water-skiing) and controlling the number of boat or activity permits.

Much of this debate hinges on the determination of lake carrying-capacities. Wood (1982) analysed the concept of carrying capacity for lake environments and links it to the concept of the recreation opportunity spectrum (ROS) (Clark & Stankey 1979). Wood argues that carrying capacity for sailing, for example, will depend on the variety of sailing opportunities being offered, as different users will have different motives for sailing and differing satisfaction levels, as noted above. In other words, he suggests that the ROS can be used to specify the factors that determine the nature of sailing opportunities and thus allow planning and management objectives for carrying capacity to be determined. These factors will include:

(1) Physical and biological factors, such as depth, wind, weeds and water quality;
(2) Social factors, such as type of sailor, type of boat and other lake users; and
(3) Management factors, such as access and safety regulations.

Wood then suggests that each factor can be monitored and management actions taken if capacity is exceeded in any one type of environment.

Lakes as Complex Destination Amalgams in Need of Management

To be considered a true tourism destination, a lake attraction needs to be complemented by support services for tourism (such as accommodation, retailing, and food and beverage), access and, ideally, a strong organisation at the destination level. Coordination of these elements is difficult at all destinations, often leading to uneven quality of provision across, say, accommodation and food and beverage. A further complicating issue is that for larger lakes, the destination management organisation may be different on various parts of the lake shore – this is an issue with the Dead Sea, for example, where responsibility is shared by Jordan, Israel and Egypt.

Integrated Lake Management

In order to maintain quality of both the lake resource and the recreation/tourism experience, management intervention is required. However, effective lake management has to recognise the distinguishing characteristics of lakes:

- Lakes are one of the best defined ecosystems on earth as they have clear-cut boundaries (Hashimoto & Barrett 1991).
- Lakes are open systems exchanging energy and materials with the environment, therefore they are vulnerable as the state of a lake is dependent on these processes of external variables or forcing functions (such as the addition of pollutants).

These characteristics demand an integrated management approach that recognise that lakes are part of wider water catchment systems (Nakamura *et al.* 1989). Management therefore has to take into account the fact that lakes are dependent not only on the natural, but also on the human activities that occur in their catchments. In fact, it is the rapid rate of socio-economic development in lake basins that has become a major threat to the quality of lake environments (Hashimoto & Barrett 1991). Management options include (Warren 1988; Massachusetts Water Resources Commission 1994):

- In-lake management – dredging or chemical treatment designed to enhance particular uses of the lake such as sport fishing.
- Watershed management – land-use planning, erosion control or access control, designed to prevent land-based problems and ideally an integral part of any lake management scheme.
- Non-intervention.

Decisions as to which of these options is appropriate will depend on the long-term costs, benefits and impacts of management. However, the integrated approach is now accepted as the most effective by taking a holistic view of the lake in order to ensure that unintended consequences do not flow from management actions. For example, management actions to reduce sediment flow into a lake may lead to the construction of shoreline structures that impact on the lake's landscape beauty. Klessig (2001) identifies the key considerations for lake management in this integrated management approach as:

(1) aesthetics;
(2) collective security (where lakes from political boundaries);
(3) cultural opportunity (community festivals);
(4) economic opportunity;
(5) educational opportunity;

(6) emotional security and spiritual dimensions (identification, natural values and memory);
(7) environmental security;
(8) individual freedom (property rights);
(9) individual security (second home owners); and
(10) recreational opportunity.

Integrated lake management can have three interpretations (Mitchell 1990a):

(1) The systematic management of the various dimensions of the lake water – ground water, surface water, quantity and quality.
(2) The management of the lake and its interactions with other systems – land and the environment.
(3) The management of the lake and its interactions with the social and economic environment – a sustainability approach.

While it is generally agreed that integrated management is the most effective approach, in practice it can be difficult to operationalise. As a result, implementation has been hesitant and sporadic (Mitchell 1990a) for two related reasons. First, the fragmentation of responsibility for the various elements of the lake – surface and ground water, for example – hinders the cooperative approach needed for integrated management. Second, boundary problems hinder integration, such as where the lake itself is managed by one agency and the surrounding wetlands are managed by a different one (Mitchell 1990b).

For large lake systems, the integrated approach does create a tension between the considerations of individual lakes and the management of the wider lake basin. Many integrated plans for these larger regions now have lake classifications to act as management templates for individual lakes within the larger basins. Commonly these are done on the basis of hydrology, depth, shoreline configuration and sensitivity to pollution and recreational use – in other words, factors that determine a lake's capacity to take use. A model for Wisconsin has developed the following classification (http://www.dnr.state.wi.us/org/water/fhp/lakes/fs1.htm):

- *Natural environment lakes/wild lakes* Here qualities such as natural beauty, wildlife and excellent water quality are protected. In other words, the management objective is special care and strictly restricted development.
- *Intermediate lakes* These lakes have reasonable levels of development but remain relatively intact as natural resources. Management objectives are to maintain present levels of use and retain existing values
- *General development lakes* These lakes already have relatively high levels of development, but due to their size and characteristics could take further development.

Lake Management for Recreation and Tourism

Overlaid across traditional management issues for lakes is their role as recreation and tourism destinations and resources. Lakes are often close to urban areas as market-oriented attractions and part of a larger recreational resource such as a park. Management of such lakes generally falls under municipal or state-level agencies (Warren 1988). Effective management of lakes is critical for the delivery of a high quality recreation and tourism experience (Warren 1988). Management approaches include:

- phasing tourism use in time;
- zoning lake use in space;
- stakeholder conflict resolution;
- lake user conflict resolution;
- codes of conduct for lake users;
- community based planning and management; and
- planning regulations

Referring to the last point in the above list, Newsome *et al.* (2002) outline simple planning regulations that can reduce the impact of tourism on lake resources. These include 'setback regulations' that determine the distance from the lakeshore to recreation/tourism development. In the USA, setbacks in wilderness and back-country environments range from 2 yd (2 m) 0.6 mile (1 km) and average 32 yd (30 m). Setbacks reduce pollutants and protect banks and fringe vegetation from damage. Setbacks also ensure that lakeshore beauty is not compromised and prevent prime-site camping pitches from limiting access to the water to other campers (Newsome *et al.* 2002).

Examples of Integrated Lake Management for Tourism

The Lake District National Park, UK

In the UK, the Lake District National Park is a unique landscape of fells and lakes, used as a resource for agriculture, mining and quarrying, and power generation, but above all it is a landscape under tremendous tourism and recreation pressure, attracting around 15 million visitors annually. Surveys suggest that it is the park's landscape that is the main attraction (Nash 1990). The human-scale landscape mosaic of lakes, tarns, fells and mountains create the appeal of the region, intimately bound up also with its personalities, history and cultural traditions – the history of the area is rich and lakeshores have many internationally significant historical and industrial archaeological sites, while more recent developments have seen Victorian villas built on the lake shorelines. The Lake

District National Park is managed by the Lake District National Park Authority, which has developed a rolling National Park Plan, which is constantly updated. The most recent version of the plan has a sophisticated and comprehensive strategy for the integrated management of the region's lakes and tarns (Lake District National Park Authority 1999).

In the Lake District, the lakes and tarns make a vital contribution to the identity and beauty of the national park. They also support an exceptional variety of aquatic plants and animals and several lakes are of national and international importance for nature conservation. Some of the lakes have a long tradition of recreational use and well-developed access and recreation facilities. Other lakes are valued for their tranquil, uncluttered and relatively undisturbed atmosphere. They provide local and regional supplies of water and a focus for national research in freshwater ecology. The objectives of the National Park Management Plan for the lakes and tarns are to (Lake District National Park Authority 1999):

- Ensure the surface waters and ground waters in the national park are of the highest possible water quality and the diversity of surface water types are maintained.
- Maintain conditions for self-sustaining populations of the full range of indigenous plants and animals associated with fresh water and wetland habitats.
- Permit the quiet enjoyment of the lakes and tarns in ways that are compatible with maintaining their character and that of the surrounding area.

These objectives could be said to conflict, and to prevent this the lakes and tarns are carefully managed for footpath use, access to the lakeshore and water surface and accommodation such as camping and caravan sites. However, such is the tourism demand for the region, that lakeshores have come under considerable development pressure (Jones 1998). The basis of the plan is to ensure that the amount of recreational use is appropriate to the character of each lake and tarn, with power boating restricted, and in congested waters, a speed limited is imposed. The plan has developed an informal lake classification scheme to guide this policy:

- lakes and tarns in quieter areas;
- lakes with nature conservation interests particularly vulnerable to recreational activity;
- other lakes.

Each of the larger lakes has a management plan (one of the earliest was for the most popular lake – Windermere (Windermere Steering Committee 1980)). These plans have been prepared in consultation with all stakeholders. The implementation of the management plan is through the planning policies of the park's local plan (Lake District National Park Authority 1998). The Local Plan has strict policies on development – for example in terms of lakeshore development the plan states that:

> Development within or on the edges of lakes and tarns will not be permitted, except for development which has a function in providing facilities associated with appropriate recreational uses ... or involves appropriate changes to existing development and which in either case cannot reasonably be located elsewhere ... (Lake District National Park Authority 1998: Policy NE7)

The Great Lakes, USA and Canada

In North America, the Great Lakes are a good example of holistic management of a lake ecosystem. The Great Lakes are an important part of the physical and cultural heritage of North America and represent the largest freshwater ecosystem on earth. The Great Lakes are used for water supply, transportation, power generation, recreation and other uses. The Great Lakes Basin is home to around one-tenth of the population of the USA and around 50% of the population of Canada.

In the USA, the Great Lakes ecosystem is managed by the Great Lakes National Programme Office, which oversees and helps all stakeholders to adopt an integrated ecosystem approach to protect, maintain and restore the integrity of the Great Lakes. The ecosystem is subject to an international management agreement between the USA and Canada, providing the basis for international approaches to the management of a shared resource. As with the Lake District National Park Plan, each lake has its own management plan (lake-wide management plans) linked by an integrated management approach. In 2002, the USA developed the Great Lakes Strategy which was designed to restore the ecosystem to one where it is safe to swim, fish and develop the aesthetics of the ecosystem. The strategy focuses on multi-lake and basin-wide environmental issues, and establishes common goals across the ecosystem and its stakeholders. The vision of the strategy is:

(1) The Great Lakes Basin is a healthy environment for wildlife and people.
(2) All Great Lakes beaches are open for swimming (by 2010, 90% of beaches will be open for 95% of the season).
(3) All Great Lakes fish are safe to eat.
(4) The Great Lakes are protected as a safe source of drinking water.

Both regions demonstrate the careful balance between overall ecosystem management and consideration for individual lakes and also the increasing priority given to recreation and tourism in integrated lake planning.

Conclusion

Lakes are significant tourism destinations in their own right and as such demand careful management and planning not only to retain their physical integrity, but also to ensure the quality of the experience of the visitor. Recreation and tourism are relatively recent uses of lakes and reservoirs, but have benefited from the increased sophistication of integrated lake management. This has, however, meant that the management of lakes has become increasingly complex as their use has diversified (Sewell 1974). As tourism and recreation have become accepted users of lakes, so has their planning and management improved, and the potential impact of visitation on the lake environment is now better understood. Despite this, many of the threats to lakes as tourism destinations come from outside the tourism system itself. These threats include climate change, decreasing water quality due to pollution and siltation due to changing agricultural practices.

In a significant international initiative in March 2003, the International Lake Committee Foundation (ILEC) delivered a call to action regarding the sustainable use of lakes. The call to action provides seven guiding principles of strategies and opportunities to provide a blueprint for the management of lakes for sustainable use (www.ilec.or.jp/wwf/index.html). The principles are:

(1) A harmonious relationship between humans and nature is essential for the sustainability of lakes.
(2) A lake drainage basin is the logical starting point for planning and management actions for sustainable lake use.
(3) A long-term, proactive approach directed to preventing the causes of lake degradation is essential.
(4) Policy development and decision-making for lake management should be based on sound science and best available information.
(5) The management of lakes for their sustainable use requires the resolution of conflicts among competing users of lake resources, taking into account the needs of present and future generations and of nature.
(6) Citizens and other stakeholders must participate meaningfully in identifying and resolving critical lake problems.
(7) Good governance, based on fairness, transparency and the empowerment of all stakeholders, is essential for sustainable lake use.

These principles are a fitting conclusion to this chapter as each principle has relevance to the role of lakes as tourism destinations and reflects the contemporary integration of recreation and tourism use of lakes with their more traditional uses.

Acknowledgements

I am grateful to Lisa Ruhanen for her help in compiling this chapter.

References

Boniface, B. and Cooper, C.P. (2001) *Worldwide Destinations: The Geography of Travel and Tourism*. Oxford: Butterworth-Heinemann.

Clark, R. and Stankey, G. (1979) *The Recreation Opportunity Spectrum: A Framework for Planning, Management and Research*. US Forest Service Technical Report PNW-98. Washington, DC: USForest Service Technical.

Collins, S. (1997) Living on top of the world. *Geographical Magazine* 69 (2), 44–5.

Cooper, C.P. Fletcher, J. Gilbert, D. Shepherd, R. and Wanhill, S. (1998) *Tourism Principles and Practice*. Harlow: Longman.

Edington, J.M. and Edington, M.A. (1986) *Ecology, Recreation and Tourism*. Cambridge: Cambridge University Press.

Fadali, E. and Shaw, W.D. (1998) Can recreation values for a lake constitute a market for banked agricultural water? *Contemporary Economic Policy* 16 (4), 433–41.

Hadwen, W.L., Arthington, A.H. and Mosisch, T.D. (2003) The impact of tourism on dune lakes on Fraser Island, Australia. *Lakes and Reservoirs: Research and Management* 8 (1), 15–26.

Hashimoto, M. and Barrett, B.F.D. (eds) (1991) *Guidelines on Lake Management Volume 2 Socio-economic Aspects of Lake Reservoir Management*. Washington, DC: United Nations Environment Program.

Jaakson, R. (1989) Recreation boating spatial patterns: Theory and management. *Leisure Sciences* 11 (1), 85–98.

Jones, C. (1998) UK national parks: Tourism in Grasmere. *Geography Review* 11 (5), 2–6.

Jorgensen, S.E. and Matsui, S. (1997) *Guidelines of Lake Management Volume 8 The World's Lake in Crisis*. Washington, DC: United Nations Environment Program.

Jorgensen, S.E. and Loffler, H. (1990) *Guidelines of Lake Management Volume 3 Lakeshore Management*. Washington, DC: United Nations Environment Program.

King, J. G. and Arnett, C.M. (1974) Effects of recreation on water quality. *Journal of the Water Pollution Control Federation* 46 (11), 2453–9.

Klessig, L.L. (2001) Lakes and society: The contribution of lakes to sustainable societies. *Lakes and Reservoirs: Research and Management* 6 (1), 95–101.

Lake District National Park Authority (1998) *Lake District National Park Local Plan*. Kendall: Lake District National Park Authority.

Lake District National Park Authority (1999) *Lake District National Park Management Plan*. Kendall: Lake District National Park Authority.

Liddle, M.J. and Scorgie, R.A. (1980) The effects of recreation on freshwater plants and animals: A review. *Biological Conservation* 17 (2), 183–206.

Massachusetts Water Resources Commission (1994) *Policy on Lake and Pond Management for the Commonwealth of Massachusetts*. Boston, MA: MWRC.

Mitchell, B. (1990a) Integrated water management. In Mitchell, B. (ed.) *Integrated Water Management* (pp. 1–21). London: Belhaven.

Mitchell, B. (1990b) Patterns and implications. In Mitchell, B. (ed.) *Integrated Water Management* (pp. 203–18). London: Belhaven.

Nakamura, M., Hashimoto, J.G. and Bauer, C. (1989) Planning for sound management of lake environments. In S.E. Jorgensen and R.A. Vollenwieder (eds)

Principles of Lake Management Volume 1 (pp. 115–40). Washington, DC: United Nations Environment Program.

Nash, J. (1990) Open for play. *Sport and Leisure* (31), 4–15.

Newsome, D., Moore, S.S.A. and Dowling, R.K. (2002) *Natural Area Tourism: Ecology, Impacts and Management*. Clevedon: Channel View Publications.

Parkes, J.G. (1974) User response to water quality and water-based recreation in the Qu'apelle Valley, Saskatchewan. In Leversedge, F.M. *Priorities in Water Management* (pp. 99–116). Victoria: Western Geographical Series 8.

Pitts, D.J. and Anderson, D.R. (1985) Recreation planning at Lake Wivenhoe. *Australian Parks and Recreation* 21 (3), 11–16.

Sewell, W.R.D. (1974) Water resources planning and policy-making: Challenges and responses. In F.M. Leversedge (ed.) *Priorities in Water Management* (pp. 259–87). Victoria: Western Geographical Series 8.

Wang, C.P. and Dawson, C.P. (2000) Recreation conflicts and compatibility between motorboat owners, personal watercraft owners and coastal landowners along New York's Great Lakes Coast. Available at www.cce.cornell.edu/seagrant/great-lakes-marinas/pwcreport.html

Warren, R. (1988) Lake maintenance brings quality recreation. *Parks and Recreation* 23 (4), 38–40.

Water Space Amenity Commission (WSAC) (1977) *The Recreational Use of Water Supply Reservoirs in England and Wales*. London: WSAC.

Windermere Steering Committee (1980) *Windermere. A Management Plan for the Lake*. Kendall: Windermere Steering Committee.

Wood, J. (1982) Lake carrying capacity. *Recreation Australia* 2 (1), 30–40.

Wood, J. and Hooy, T. (1982) Planning for quality sailing experiences on Canberra's lakes. *Australian Parks and Recreation* 18 (5), 14–21.

World Tourism Organisation (2002) Unpublished draft document presented at the WTO joint meeting of the Destination Council and the Education Council. Madrid: World Tourism Organisation.

Part 2: Historical Dimensions and Cultural Values

Chapter 3
Natural Heritage and the Maintenance of Iconic Stature: Crater Lake, Oregon, USA

STEPHEN R. MARK

National parks have become probably the most recognised and globally pervasive way of perpetuating natural heritage over the past century or so. Lakes are sometimes included within the boundaries of these protected areas, but only a few of them serve as central features. The namesake of one national park in North America attained 'iconic' status during the late 19th century, although subsequent cultural perceptions of Crater Lake have not remained static. Those perceptions have historical roots, but are layered while changes in hierarchical positioning occur within a broader national system of such parks. Identifying three stages in how one culture has seen Crater Lake are thus important to any discussion of how park managers have responded to public attitudes, but also to understand the park's changing place in a scenic hierarchy of other protected areas.

Introduction

At 1943 ft (592 m), Crater Lake is the deepest freshwater body in the United States and holds seventh place in the world. As a defining characteristic (one sufficient to merit national park status as far back as 1902), however, depth defers to beauty as the lake's most prominent feature. Its spectacular setting is derived from a cataclysmic volcanic eruption that occurred at the site some 7700 years ago, one that moved almost 8 sq. miles (20 km^2) of material from Mount Mazama and made this event the region's most recognisable stratigraphic time marker. In the eruption's wake, the mountain collapsed along a ring fracture zone to leave an oversized crater called a caldera. So-called 'back flows' sealed the sides and bottom sufficiently in order for thermally heated water and precipitation to collect over several hundred years to form the early lake. Subsequent volcanic activity produced platforms, domes and cones. All but one cinder cone (called Wizard Island) and remnants of an earlier cone (known as Phantom Ship) remain submerged within a caldera whose jagged cliffs rise up to 2000 ft (600 m) from the lake surface. A number of nearby peaks can be seen as background to form an interesting contrast between the violence that created the larger Cascade Range and the placid beauty suggested by blue water, open sky and a verdant coniferous forest. To geologists, Crater Lake represents a model for how small calderas evolve, with the lake representing merely one state in the mountain's development – one that will end abruptly when there is renewed volcanic activity.

Scientific study of Crater Lake began in 1883, when scientists with the US Geological Survey made their first reconnaissance in that part of the Cascades. Europeans had first seen the lake three decades earlier, although the earliest mention of its possibilities as an attraction for tourists had to wait for road-builders from the US Army in 1865. The first leisure tours in the region became feasible with the arrival of a transcontinental rail link during the mid-1880s, so it is no coincidence that the first petition to the United States Congress for protecting Crater Lake as a national park came in 1885. Parsimonious members in both houses (many of whom believed that national parks represented a continual drain on the federal treasury holding few benefits for most of their constituents) furnished much of the reason why it took until 1902 to establish Crater Lake National Park. The interest of President Theodore Roosevelt and some of his key advisers proved crucial to the passage of a bill establishing the park, given that fewer than 1000 visitors came to Crater Lake in any year prior to 1902.

Over the following 75 years, the annual number of visitors grew to some 617,000 by 1977 (Table 3.1). It has since remained relatively constant at a mean figure of half a million, despite dramatic population jumps in three states that serve as the residence for more than 80% of park visitors: Oregon, California and Washington. One reason is the draw of a central feature (Crater Lake) to the exclusion of anything else in the park, such that only 1% of all visitors ever venture away from roads (one circles the lake) and into the back-country. A second factor is the decidedly seasonal nature of the visits, given how fewer than 10% of the total come during the six months of winter, with most opting for July and August. Visitors generally see Crater Lake on their way to other destinations in southern Oregon or northern California, although the park is located some distance from the main north–south highway corridor. Crater Lake National Park is therefore largely a regional attraction, with the three west coast states accounting for more than 70% of all visitors, while those originating from outside the United States are only 2% of the total (United States Department of the Interior – National Park Service (USDI-NPS) 1997).

The rise of competing attractions such as river rafting or even casino gambling since the 1970s might furnish an additional reason why the annual visitation to Crater Lake has not grown over the past quarter-century. Touring the national parks in family groups by car has declined in general, at least among the working population, as Americans enjoy the least amount of time away from their jobs among all industrialised nations. Nevertheless, population growth in the region has resulted in expanding visitor numbers where other national parks are closer than Crater Lake is to large metropolitan centres. Another factor to be considered is the relatively passive type of visitor drawn to Crater Lake, where sightseeing from the rim greatly exceeds all other activities. Boat tours attract only 5% of visitors (some 25,000) in the best years, mainly because the lake can only be reached by one trail that descends 700 ft (213 m) over the course of a mile to the shore. Many visitors thus have little interaction with the water in comparison to other lakes where road access culminates at a boat ramp. Instead, they see Crater Lake as distant and untouchable, only visible from various positions on the rim. Wizard Island and Phantom Ship serve as distinct or shifting reference points, depending on where those visitors stop on the road circuit around Crater Lake.

Crater Lake's Position in a Scenic Hierarchy

Annual visitation and the structural or historical factors affecting it provides only part of the answer to explain why some scenic areas receive more funding than others in the same region, or across a system containing what appear to be equally outstanding features representing natural heritage. In the American context, the place of Crater Lake in a wider 'scenic hierarchy' must be understood with reference to which jurisdiction administers these specially designated areas. As the name implies, a federal authority manages national parks. Since their inception, the parks have been lodged in the US Department of the Interior, but from 1916 onward have been managed directly by a bureau called the National Park Service (NPS). At the time of writing, the NPS has charge of 388 distinct units divided into roughly 20 categories. National parks, presently 53 in number, are only one of the categories, although larger staff sizes and operating budgets tend to correlate with this designation.

It should be noted, however, that popular scenic areas are not the sole property of the NPS in western states such as Oregon. The US Forest Service (USFS), a bureau in the US Department of Agriculture, manages far more land in Oregon than the NPS. A vast number of lakes, many of which are in legally designated federal wilderness areas, are in USFS jurisdiction and visited by hikers, fishermen, hunters and horse users each summer. Other agencies in the USDI, such as the US Fish and

Table 3.1 Annual visitation to Crater Lake National Park

Year	Number of visitors	Year	Number of visitors
1902	unavailable (author's estimate 1000)	1954	370,554
1903	unavailable (author's estimate 1500)	1955	343,839
1904	estimated 1,500	1956	359,840
1905	estimated 1,400	1957	330,398
1906	estimated 1,800	1958	333,893
1907	estimated 2,600	1959	340,989
1908	5,275	1960	330,398
1909	4,171	1961	415,567
1910	estimated 5,000	1962	592,124
1911	estimated 4,500	1963	475,684
1912	5,235	1964	497,057
1913	6,253	1965	480,478
1914	7,096	1966	552,531
1915	11,371	1967	499,357
1916	12,265	1968	578,271
1917	11,645	1969	544,932
1918	13,321	1970	534,962
1919	16,645	1971	557,958
1920	20,135	1972	594,343
1921	28,617	1973	539,708
1922	33,016	1974	525,030
1923	52,017	1975	427,252
1924	64,312	1976	606,636
1925	65,018	1977	617,479
1926	86,019	1978	580,061
1927	82,354	1979	446,397
1928	113,323	1980	484,256
1929	128,435	1981	536,719
1930	157,693	1982	484,283
1931	170,284	1983	429,586
1932	90,512	1984	499,945
1933	not available	1985	427,927
1934	116,699	1986	427,716
1935	107,701	1987	492,581
1936	180,382	1988	468,994

Table 1.1 continued

Year	Number of visitors	Year	Number of visitors
1937	202,403	1989	454,737
1938	190,699	1990	454,253
1939	225,101	1991	525,441
1940	252,482	1992	511,500
1941	273,564	1993	419,914
1942	100,079	1994	497,423
1943	28,637	1995	542,611
1944	42,385	1996	526,557
1945	77,864*	1997	501,148
1946	188,794	1998	521,465
1947	289,681	1999	476,168
1948	243,533	2000	476,483
1949	255,610	2001	506,973
1950	328,041	2002	506,219
1951	306,568	2003	528,782
1952	323,410	2004	466,665
1953	332,835		

Note: * 1945 and preceding years based on travel year ending in September.

Wildlife Service and the Bureau of Land Management, also manage scenic areas in Oregon, although their share of recreational visits compared to lands administered by the Forest Service is comparatively small. Despite only having a small portion of Oregon's land base (95,129 acres (384,968 ha) out of a statewide total of 62 million acres (25 million ha)) state parks there receive the fifth heaviest use of any in the nation. This figure (currently 37 million visitors per year) is far out of proportion to the state's population of 3.4 million (which is roughly 1% of the national total), although California, with some 35 million residents, is to the south and Washington (5.9 million) lies to the north. The usage figure makes Oregon state parks second in park visitors per acre (0.4 ha) (413 as opposed to the national average of 84), with camping provided so that the state parks allow 2.4 million overnight stays per year, enough for eighth place in the United States.

Crater Lake is the only national park in Oregon, whereas California contains six and Washington has three. Its annual operating budget of US$4 million covers equipment (winter snow removal consumes about one-sixth of this figure) and salaries for roughly 55 full-time staff and almost that many seasonal workers, along with other expenses such as utilities. Even when project funds ('soft money') is factored in, Crater Lake falls next to last among the ten national parks located in west coast states (Table 3.2). It is placed nowhere near Yosemite National Park in California, whose US $23 million budget in 2003 was derived to some extent from its 3.4 million visitors that year, but also from the demands (some political) imposed by proximity to millions of constituents. Although Yosemite boasts the largest budget among areas designated as national parks on the west coast, other kinds of units administered by the NPS can dictate greater outlays of federal funds than does Crater Lake. Lake Roosevelt National Recreation Area (most of which is the land surrounding a water impoundment from Grand

Table 3.2 US national parks located in the three west coast states, 2003

Park (location)	Size (acres)	Visitation (2003)	Base budget* 2003 (US$ million)	(US$ million) 2005
Channel Islands NP (Calif.)	249,561	413,000	4.96	5.83
Crater Lake NP (Ore.)	18,224	451,000	4.01	4.21
Death Valley NP (Calif.)	3,372,402	854,000	6.90	6.96
Lassen Volcanic NP (Calif.)	106,372	407,000	3.73	3.88
Mount Rainier NP (Wash.)	235,625	1,857,000	9.10	9.87
North Cascades NP (Wash.)	684,302	430,000	5.60	6.06
Olympic NP (Wash.)	922,651	3,000,000	10.3	10.36
Redwood NP (Calif.)	112,613	406,000	7.40	7.26
Sequoia/Kings Canyon (Calif.)	864,411	1,552,000	13.0	13.31
Yosemite NP (Calif.)	761,266	3,388,000	23.1	23.43

* Includes only the share of operating funds appropriated by Congress.
Note: Visits to all NPS units totalled 266 million for 2003.

Coulee Dam in Washington), for example, exceeded what Crater Lake received in funding by the early 1990s.

As an index to Crater Lake's position in a scenic hierarchy, its operating budget might serve as a leading indicator if such funding solely reflected an area's relevance to the regional economy and if the NPS administered all areas where scenic tourism contributed to that economy. In Oregon, however, the NPS is a comparative stranger to the state's residents in that it manages only four areas totalling some 195,000 acres (78,912 ha) (Crater Lake National Park comprises more than 90% of that figure) with most being somewhat removed from major population centres such as Portland, Salem and Eugene. (The other three are John Day Fossil Beds National Monument, Lewis & Clark National Historical Park (formerly Fort Clatsop National Memorial) and Oregon Caves National Monument). Several national forests lie much closer to these cities, such that Oregonians have largely looked to the USFS to supply them with camp-grounds and other recreational facilities on federal land (which covers about half of the state) since the 1920s. Visitation to some USFS recreational sites near Portland, such as those in the Columbia River Gorge and on Mount Hood, dwarfs that of Crater Lake. One Forest Service site in particular, Multnomah Falls, is the single most heavily used scenic area in Oregon, mainly due to its location on an interstate highway 30 miles (50 km) east of Portland.

Even if much of the recreational use in Oregon lies to the north of Crater Lake in the national forests, it is probably the most intensely centred on the state's coastline. The USFS maintains a presence there too, being charged with administering one national recreation area (the Oregon Dunes), along with several coastal waysides and some camp-grounds. Visitation on the Oregon coast, however, is mainly focused on a string of state parks that provide access to roughly 300 miles (500 km) of public beach. Only in aggregate does the number of visitors at the coast greatly exceed those coming to Crater Lake, since none of the individual state parks are large enough to draw half a million people a year by themselves without a major north–south highway (US 101) connecting them. This is perhaps why Crater Lake garnered enough support to represent Oregon on a 25 cent federal coin that makes its debut in 2005 as part of a 'state' series issued by the US mint. The Crater Lake image outpaced three other designs depicting a jumping salmon (an increasingly rare species throughout the state), a covered wagon (which reflects a heritage of being at the end of the Oregon Trail) and Mount Hood, the state's highest peak. Although Crater Lake hardly dominates tourism in Oregon (currently the second largest industry in the state), its stature as the lone national park provided a common rallying point among selectors – some of whom hailed the image as 'iconic' instead of being merely representational.

Stages in Cultural Perception: Crater Lake as Sublime

Icons, at least in the traditional sense, were to be 'written' not drawn because the artist was moved by the Holy Spirit. Although the Greek *eikon* meant 'image', artists created icons for evangelism and outreach rather than patrons and museums, so the popular meaning of icon as advertising symbol belies the original intent of inspiring the viewer to an awareness of divine presence. The traditional icons remained two-dimensional to avoid realism and thus accusations of idolatry, although the practices of those who supported icons had their roots in polytheism.

Like icons, the sublime also has its roots in religious experience associated with Christianity. Their time frames differ, in that religious icons began to be produced in the 5th century, whereas the sublime came about as a reaction to transition towards industrial capitalism starting in the 18th century. Artists in particular sought to resist a loss of status in the new industrial order of society founded on contract and the status provided by access to capital and the means of production. Nature, a word whose meaning was once confined to describing people, expanded its scope to include the world beyond the immediate grasp of human beings. To many thinkers and artists, such a world of nature was held to be indicative of truth, whereas a social order that regarded people as instruments of mechanical production had to be false. Closest to truth and indicative of infinity, grandeur and gloom lay the sublime – which also extended nature to produce wonder in the individual and even inspiration. It could be found in nature, but in wild and uncultivated places supposedly untouched by industrial consumption, in areas so dramatic that there might be direct contact with a divine creator (Cosgrove 1998).

Majestic natural forms became central to the sublime experience, one that allowed even the new middle classes to participate eventually since they could potentially travel to find suitable evocative scenery. Particularly important was the sense of sight, one critical for appreciating wild nature in person and through landscape paintings created to enhance and transmit the effects of craggy recesses, storms and the minuteness of human presence in such settings. Artistic licence notwithstanding, these images drove a small number of early and mid 19th century tourists to seek the sublime in new localities beyond Europe, to the North American West and elsewhere (Ross 1998).

The breadth of acceptance for the sublime and its associated aesthetic (which conditioned people to see nature in such a uniform way) accounts for the reason why the jump between the advent of public parks in rapidly growing cities and reserving places such as Yosemite Valley in relatively inaccessible mountain ranges could take place so quickly. Central Park in New York City opened in 1857 in response to fear of social upheaval that might accompany a densely packed urban area where industrialisation segregated immigrants and natives alike by social class. The first public park in the United States was intended for these classes to mix, while all enjoyed the benefits of exercise and contact with nature. It followed precedents established in England and Germany, where public spaces were developed to imitate pastoral nature in accordance with how private estates had followed the rules of artistic composition in an earlier fusion of garden and deer park. By 1864, the US Congress gave the federal land surrounding Yosemite Valley and a grove of giant sequoias called the Mariposa Big Trees to the State of California, stipulating that the grant was to be held for 'public use, resort and recreation' (Act of 30 June 1864; for more detail see Runte 1990).

The Yosemite grant and contemporary landscape paintings depicting the valley helped forge a link between sublime nature and American national identity, something furthered in 1872 when Congress passed legislation creating a public park on the remote Yellowstone Plateau, a reservation so large that parts of it were shared by the future states of Wyoming, Montana and Idaho. Other such parks followed, albeit slowly, as Congress seemed reluctant to appropriate money for managing such reserves, given how few of their constituents were willing to ride the railroads and stages of the time in order to reach them. Over the next three decades, just five more 'national' parks were approved before Crater Lake joined the list in 1902. In some respects, it was remarkable that any such legislation had been passed, as Congress approved a number of laws during that period aimed at disposing the, as yet unclaimed, public domain to settlers, miners and railroads. Nevertheless, the sublime's grip on American consciousness proved great enough to establish a precedent, although it had also been helped by scientific discoveries of seemingly harmoniously 'laws' governing how the earth formed. This gave tourists and those who read their accounts a way of seeing spectacular landscapes not only as curiosities, but also as geological features that were discernible through an understanding of natural law.

Fuelled by both the sublime and potential rewards associated with lucrative mineral deposits, geology emerged as the leading science of the 19th century. In stressing the uniformity between ancient and modern processes, geology attempted to explain the story contained in layers of rock – one that many people believed showed how the natural law imposed by an omnipotent creator allowed progress and unity to manifest divine wisdom. Geology could thus allow one to 'read from the book of nature', but also provide insight on the meaning behind sublime spectacles such as Yosemite or Crater Lake.

Without specialised training, many tourists contented themselves with natural history. This pursuit depended on the notion that anyone's knowledge could be broadened by travel and study, particularly that aimed at nature. They collected in the belief that creation could be systematised, using classification schemes (such as the Linnaean, for example) intended to illuminate the natural relations among rocks, animals and plants. Searching for beauty and significance through natural history allowed the tourist a combination of personal discovery, self-improvement, piety, industry and recreation. Essayists fuelled this interest by often presenting facts derived from geology with more romantic depictions of wild life, indigenous peoples and activities such as mountaineering (Pomeroy 1990).

Not all 19th-century tourists indulged in such high-minded endeavours such as geology and natural history, however. Some came simply to be seen, moving on to new sights on various 'circuits' established to approximate earlier 'grand tours' conducted in Europe. In any event, promotion of areas destined to become national parks followed a predictable sequence. First came discovery (always by white men, but never non-English speaking immigrants) by one who could not wish to claim it for themselves. Next was the 'unveiling' as a curiosity, presented first in newspapers and then guidebooks written to assist those in pursuit of economic opportunity through relocation, with some sublime locales added to spice the account. Veneration of those landscapes selected as sublime (but otherwise worthless to miners and settlers) took place as the transportation network led by railroads began to integrate the region with the larger national market economy. Mapping and identification of what could be retained as pristine or 'untouched' (not mined, logged or settled as private property) followed the onset of veneration, a process that could lead to formal establishment of a park.

Indigenous people living in the surrounding river basins knew about Crater Lake and had it integrated into their stories about the world well before a group of would-be miners 'discovered' the lake in 1853. The miners could not publicise their find in print, owing to the fact that the first newspaper in southern Oregon did not make its debut for another three years. That organ eventually carried the first written account about Crater Lake, but not until 1862 when another party of miners travelling from goldfields in eastern Oregon came across the vast expanse of water while attempting to cross the Cascade Range. The construction of a wagon road through those mountains in 1865 made the lake into a curiosity, as an account printed in the closest paper (one in Jacksonville, Oregon) reached Philadelphia within a week. Titled *The Sunken Lake*, this piece read like many others of the time publicising new scenic wonders in the West:

> This lake rivals the famous valley of Sinbad, the sailor. It is thought to average two thousand feet down to the water, all around. The walls are almost perpendicular, running down into the water and

leaving no beach. The depth of the water is unknown and its surface is smooth and unruffled and it lies so far below the surface of the mountain that the air currents do not affect it ... No living man ever has and probably never will be able to reach the water's edge. It lies silent, still and mysterious, in the bosom of the 'everlasting hills,' like a huge well, scooped out by the hand of the giant geni of the mountain, in unknown ages gone by and around it the primeval forests watch and ward are keeping. The visiting party fired a rifle several times into the water, at an angle of forty-five degrees and were able to note several seconds of time from the report of the gun until the ball struck the water. Such seems incredible, but is vouched for by some of our most reliable citizens. The lake is certainly a most remarkable curiosity. (*Inquirer*, 17 August 1865)

Only when the first tourists (in this sense leisure travellers who did not reside in southern Oregon) visited Crater Lake in the 1870s and 1880s did it join the more select company of the sublime and was labelled as such. Crater Lake could thus be compared with better-known company such as Lake Tahoe, Yosemite Valley and Niagara Falls. The comparison with Lake Tahoe was first made in the Ashland (Oregon) Tidings of 21 September 1877. After describing Crater Lake as a 'sublime and wonderful reservoir', Joseph Le Conte, a geology professor at the University of California, wrote in 1885 that Yosemite and Crater Lake could not be compared, as each had sublime glories, although 'there is but one Crater Lake'. John Wesley Powell, explorer and head of the US Geological Survey, favourably compared Crater Lake with Niagara Falls in 1888, writing 'there are probably not many natural objects in the world which impress the average spectator with so deep a sense of the beauty and majesty of nature.'

Although the Oregon Legislature characterised Crater Lake as sublime in their petition to have it set aside as a public park that went to Congress in late 1885, supporters liked the fact that trained scientists from the newly created US Geological Survey (USGS) might lend credence to such a proposal. Even before the scientists launched their expedition to ascertain the lake's depth and origin, one of them borrowed from the vocabulary of nature poets and others who celebrated the sublime in making his case to a national audience:

It is difficult to compare this scene [on Crater Lake] with any other in the world, for there is none that sufficiently resembles it; but in a general way, it may be said that it is of the same order of impressiveness and beauty as the Yosemite Valley. It was touching to see the worth but untutored people, who had ridden a hundred miles in freight wagons in order to behold it, vainly striving to keep back tears as they poured forth their exclamations of wonder and joy akin to pain. Nor was it less so to see so cultivated and learned a man as my companion hardly able to command himself to speak with his customary calmness. To the geologist this remarkable feature is not less impressive than it is to the lover of the beautiful ... (Dutton 1886)

Congress failed to pass legislation establishing the park at that time, even after the USGS measured Crater Lake to be the deepest freshwater body in the United States. Promoters such as William Gladstone Steel, a man who had been working as a postman in Portland, but then devoted much of his life to Crater Lake after visiting in 1885, took up the cause in earnest. Steel and several others produced a steady stream of articles about the lake, often referencing its sublimity and making occasional references to better known localities such as Italy's Lake Como, while providing details about what tourists should see on their hundred-mile journey by wagon from the nearest train station. Steel's promotional efforts tried to capitalise on how the lake might now be better accessed, with a rail line reaching southern Oregon from Portland in 1884 and a connection to California in operation three years later. Steel demonstrated in 1888 that the railroad could be used to facilitate tours that included Crater Lake and Oregon Caves (a future national monument), provided that visitors could hire guides and gear to travel the wagon roads connecting both attractions with the main north–south rail corridor (Steel 1886, 1888a, 1888b). Interestingly, the first printed mention of tourists in relation to Crater Lake appeared in the article 'Crater Lake' by L.W. Hammond which was in a Portland literary magazine called *Pacific Slope* on 7 October 1874. The party visited a week or so before the first photographs of Crater Lake were taken by a Jacksonville photographer called Peter Britt. Steel used the photographs in his park establishment campaign.

Although Steel proved that a rail and wagon tour was possible, even planting fish in Crater Lake as an inducement to those who liked fishing, his efforts on behalf of national park designation did not bear fruit during the 1890s. During this decade, roughly a thousand visitors per year came to the lake, although that figure could be adjusted upward during years such as 1896. Over that summer Steel brought scientists to the lake, then hosted a large gathering of climbers from a mountaineering club (the Mazamas) he had started in Portland and finally accompanied a forestry commission appointed by the President, all as a way of publicising the park proposal. Enactment by Congress in the first half of 1902 only came when the new president, Theodore Roosevelt, personally intervened to save what many considered to be a hopeless bill. Even then, assurance had to be given to key Congressmen that the park did not include mineral deposits or lands suitable for farming. One of the USGS scientists made sure of this by making the authorised boundaries match those in a map he produced as part of writing a monograph on the geology of Crater Lake and its related volcanic features (Diller 1902).

Although Crater Lake National Park became an official designation on Oregon maps as of 1902, appropriations from Congress over the next five years were minimal in comparison to older and larger parks such as Yosemite, Yellowstone and Mount Rainier. Annual visitation in that period reached a high of only 2600 in 1907, mainly because road access to Crater Lake remained primitive, while facility infrastructure (aside from some tents in two places) was nonexistent. Writers such as Joaquin Miller liked it that way, since he eschewed Steel's plan for a great hotel and carriage road around the lake. His Crater Lake (or 'Sea of Silence', as Miller called it) should not be surrendered to hotels and railroads as some other parks such as Yellowstone had been (Miller 1904: 404).

Not that Miller's reaction could keep any plan at bay, as Steel secured the privilege of being the park concessionaire in order to offer rides on the lake in fuel-powered boats, beginning in 1907 and then started building a lodge two years later. Annual visitation remained at an average of 5000 per year from 1908 to 1913, mainly because the road network in southern Oregon had changed only a little from the wagon roads of the 1890s. Visitors during that period came to Crater Lake drawn by news articles, celebrity testimonials (the author Jack London appeared in 1911) and commercial photography which adopted the same conventions as landscape painting but could be produced for a mass market. Beggarly appropriations meant that road improvements took time since the park staff remained at two. Even when Steel and his friends secured enough funding for the US Army Corps of Engineers to begin building a road around Crater Lake in 1913, it was hitched to a much larger appropriation for roads in Yellowstone. This funding remained mired at levels so low that roadwork (which continued over the next five summers) never proceeded beyond the grading phase of construction.

As an 'independent' national park, Crater Lake languished on the lower end of funding for the parks, at least in terms of how Congressmen perceived its ability to resonate for a large number of voters (Tables 3.3 and 3.4 indicate the appropriations for 1910 and 1915 respectively). Its supporters, like national park advocates generally, begin in 1912 to rally support for a central bureau charged with managing and promoting the parks. They succeeded in obtaining congressional authorisation for such an agency in 1916, one directed to protect the sublime qualities of Crater Lake and other national parks from impairment, yet also responsible for directing how facilities could be better developed. As they had in many other areas, Americans thought that the way to have the park administration reflect their notions of efficiency and scenic nationalism was through making them into a system.

Table 3.3 Visitation and funding for selected US national parks, 1910

Park (location)	Date established	Visitation	Budget $US
Yosemite (Calif.)	1864 (1890)	13,619	30,000
Yellowstone (Wyo.)	1872	19,575	8,000
Sequoia (Calif.)	1890	2,407	15,550
General Grant (Calif.)*	1890	1,178	2,000
Mount Rainier (Wash.)	1899	8,000	3,000
Crater Lake (Ore.)	1902	5,000	3,000
Wind Cave (S.D.)	1904	3,387	8,500
Mesa Verde (Colo.)	1906	190	7,500
Glacier (Mont.)	1910	4,000	15,000

Note: * Incorporated within Kings Canyon NP when established in 1940. Two other national parks (Platt, Sullys Hill) not included due to delisting in the 1920s.

Table 3.4 Visitation and funding for selected US national parks, 1915

Park (location)	Date established	Visitation	Budget $US
Yosemite (Calif.)	1864 (1890)	33,452	100,000
Yellowstone (Wyo.)	1872	51,895	8,500
Sequoia (Calif.)	1890	7,647	15,550
General Grant (Calif.)	1890	10,523	2,000
Mount Rainier (Wash.)	1899	35,166	51,000
Crater Lake (Ore.)	1902	11,371	8,040
Wind Cave (S.D.)	1904	2,817	2,500
Mesa Verde (Colo.)	1906	1,000	10,000
Glacier (Mont.)	1910	14,265	75,000
Rocky Mountain (Colo.)	1915	31,000	3,000

A Second Layer of Systemisation

While integrating Crater Lake into a system of national parks might be viewed as a necessary precondition for mass access, it also had a profound effect on visitor perception and the tourist experience. The quest for efficiency is part of the industrial process, yet it affected those park areas considered sublime – once a reaction against a mechanistic social order. To some extent, the idea of establishing national parks had always hinged on the impulse to equate the sublime with glorifying the industrial nation state. Making Crater Lake and other 'crown jewels' good for business went hand in hand with a railroad slogan that urged citizens to 'See America First'. It followed that a new government bureau, the National Park Service, could be both regulator (of concessionaires in much the same way that the US Forest Service regulated commodity development on the national forests) and also partner in supplying what businessmen could not – such as building trails, fighting wildfires and collecting park entrance fees.

The first concern of the NPS, however, was agency survival. Its leaders needed to build a distinct identity from the USFS that Congress created in 1905. As a functional rival to the NPS, the USFS controlled some lands many thought should be in national parks (the Grand Canyon, for instance) and had embarked on a programme of developing some lands for recreation by the time Congress created the NPS in 1916. NPS officials knew they had to promote the existing parks as a system and then equate their bureau with what it managed in the public mind.

A 'National Park Portfolio' manifested an idea borrowed from the world of advertising, in that designers used short essays accompanied by professional photography to create a standard look. At first financed privately (the US Railroad Administration and the National Geographic Society were key donors), the portfolio soon became a government publication that lasted until the 1930s. At that point park circulars were expanded, yet retained their uniform look and purpose as promotional devices. These had evolved into brochures by the 1950s, but followed the same graphic specifications in order to reinforce a collective identity. All were generated by the NPS, which had help from numerous private publications in promoting visitation to the parks before the Second World War. Nevertheless, the tourism and recreational significance of Crater Lake in the national park system arguably diminished as the park system grew from 17 parks and 22 national monuments in 1917 to one containing 22 national parks and units with numerous other designations by 1934. That year, in an effort to promote visitation to the parks during the Great Depression, the United States Post Office issued ten postage stamps in a set depicting individual parks. Crater Lake had a place among them, to some extent because it still occupied tenth place in size, visitation and appropriations, as indicated in Table 3.5.

What directly stimulated growth in the national park system between the world wars (and indeed the creation of the NPS itself), was the tremendous growth in North American vehicle ownership that took place after 1900. In the United States it had grown from 8000 at the turn of the 20th century to 40 million by 1930 and continued upward

Table 3.5 Size, visitation, and budget for US national parks, 1934

Park (location)	Sq. miles	Visitation	Budget US$
Acadia (Maine)	18.53	238,000	55,000
Bryce Canyon (Utah)	55.06	34,140	14,000
Carlsbad Caverns (N.M.)	1.12	61,474	68,000
Crater Lake (Ore.)	250.52	110,00	63,000
Glacier (Mont.)	1533.88	53,202	201,000
Grand Canyon (Ariz.)	1009.68	121,267	136,000
Grand Teton (Wyo.)	150.00	40,000	20,000
Great Smokies (Tenn/N.C.)	465.18	300,000	28,000
Hawaii (Hawaii)	245.00	48,079	140,000
Hot Spring (Ark.)	1.45	201,762	83,000
Lassen Volcanic (Calif.)	163.32	41,723	28,000
Mesa Verde (Colo.)	80.21	15,760	53,000
Mount McKinley (Alaska)	3030.46	357	28,000
Mount Rainier (Wash.)	377.00	216,065	144,000
Rocky Mountain (Colo.)	405.33	282,980	98,000
Sequoia (Calif.)	604.00	131,398	113,000
Wind Cave (S.D.)	19.06	12,539	18,000
Yellowstone (Wyo.)	3426.00	157,624	466,000
Yosemite (Calif.)	1162.43	498,289	335,000
Zion (Utah)	148.26	51,650	47,000

Note: Parks shown in italics were represented in the set of ten postage stamps.

to 200 million over the next seven decades (Havlick 2002). This triggered more funding for roads in general (initially at the behest of promoters who called themselves the 'Good Roads Movement'), especially when states such as Oregon began to exercise centralised control over their highways after 1917. Prior to that time road construction and maintenance largely depended on county taxes, but the prospect of better state and federal funding was slowly realised. Oregon became the first state to levy a fuel tax for its highways in 1925, the same year when legislation authorising large increases in federal aid for roads passed Congress.

The NPS, meanwhile, promoted betterment of a 'National Park to Park Highway' starting in 1918, a grand loop that connected existing roads between the parks as a way to promote a system. The agency had no direct control over such a project, but worked instead to obtain funds for roads within the parks as a way to accommodate increased numbers of visitors who toured in their own vehicle. At Crater Lake, for example, work on park roads received a giant shot in the arm when Congress passed the federal aid bill and continued until American entry into the Second World War. Consequently, the narrow and unsurfaced roads built by the Army Corps of Engineers in the park during the First World War period became paved highways suited to vehicles that could maintain speeds of 50 miles per hour or more after 1928. A new 'Rim Drive', intended to replace the rough circuit road around Crater Lake, received the greatest emphasis among all the highway projects in the park. Not only did it allow the central feature to be better seen from various vantage points, but its designers intended that Rim Drive provide visitors with the means to learn about the park's geology and natural history.

One of the key ways in which the NPS tried to separate itself from the agency's main rival, the USFS, was by providing nature guides who could 'interpret' the national parks to visitors through guided hikes, scheduled talks and specially prepared literature. These 'naturalists' initially consisted of professors and graduate students who were generally acknowledged to have begun their work at Yosemite in 1920. NPS officials in Washington, DC agreed, with several leading scientists of the time declaring that the national parks represented a 'superuniversity of nature', one perfect for helping adults to read nature's 'book' since the parks contained the most spectacular landscapes in the country. An educational programme had been established at Crater Lake in 1926, quickly growing in size and scope, such that naturalists boasted that they made personal contact with most visitors, even reaching about 60% of the 202,000 who came to the park in 1937.

Naturalists were so successful in reaching visitors that they made their ranger uniform (one based on army dress of the late 19th century) into something of a ready identifier as to how national parks differed from the outside world. Another way the NPS achieved this distinction came through its programme of building facilities to blend with their natural surroundings. Such 'rustic architecture' received substantial funding through work relief programmes that dominated the construction programmes in national parks during the 1930s. Structures and facilities at Crater Lake National Park are still dominated by this influence, something said by architectural historians to constitute one of the best expressions anywhere on how to design with nature. Perhaps the most outstanding visitor facility built during this period was the Sinnott Memorial, an 'observation station' situated inside the caldera, where visitors could hear a short lecture on how Crater Lake was formed, along with orientation about where to go in the park (Mark 2002).

As a starting point, a circuit around Rim Drive, the Sinnott Memorial was hailed when it opened in 1931 as a great boon to education in the parks, partly because it did not attempt to overwhelm visitors, as some museums in other parks had, with exhibits and displays. The building blended almost perfectly with the walls surrounding Crater Lake, but did not inspire very many imitations elsewhere. A sometimes hefty winter snowfall confined the access by trail to high summer and the battered stone masonry often proved too difficult and expensive to replicate in other places, especially after the Second World War when visitation soared in state and national parks. By that time the NPS had begun to build 'visitor centres' instead of museums or facilities with an open parapet such as the Sinnott Memorial. Visitor centres incorporated exhibits and talks by naturalists, but were aimed at handling larger crowds in a predictable and standardised way, much in keeping with modern design tenets that dominated the building practices of the 1950s and 1960s. Designs for park facilities during that period had less emphasis on being 'rustic' (where the site was said to dictate the appearance of structures) and more with simply accommodating larger numbers of visitors than ever before – especially when the cheap labour available through Depression work-relief programmes had disappeared.

By 1960, the number of visitors to Crater Lake National Park had surpassed 400,000 annually, yet the NPS built no visitor centre there. Renovation of existing facilities seemed to suffice for the orientation function, while the agency made only modest upgrades to its infrastructure, such as expanded camp-grounds, a new trail to the lake, more staff housing and a few roadside exhibit panels. It resisted several proposals made during the 1950s (one was advanced by a Congressman from Oregon) to build an aerial tramway or gondola from the rim to Wizard Island, just as NPS officials opposed earlier (and somewhat more chimerical) schemes to construct a lift and then a tunnel to the lakeshore. In looking forward to celebrating the bureau's 50th year of existence in 1966, the NPS could confidently turn back what staff and outside supporters saw as a threat to visitor experience and what came to be called 'park resources'.

Although the agency had hardly been consistent in building winter sports facilities like ski lifts in some parks and not others, these did not impinge on central features such as Crater Lake. Park officials could point to the obvious visual impact that a gondola would surely have on what visitors saw from the rim; at any rate, tramways and ski lifts existed at several national forest recreational sites in Oregon and there seemed to be no need for them in the state's only national park. Crater Lake had never really been a destination resort and by the 1960s most people visited for only a couple of hours on their way to a national forest campsite or motel located in a nearby town. Camping and lodge accommodation persisted in the park, but were no longer the only such amenities in southern Oregon as they had been before the early 1920s. Statewide, tourism emerged as the third largest industry by 1950, with Crater Lake having long ceased to occupy the dominant place it once enjoyed in the early days of Oregon's tourist economy. A tram or other such gimmick made no sense from a revenue standpoint, let alone seemed worth the risk of destroying the dignity of what visitors had come to expect from a visit to the lake (Smith 1959). Oregon's other Congressman, Al Ullman, opposed Porter over the tramway (Smith 1960). However, most NPS officials and some visitors enjoyed their opposition aired publicly, even in the face of the park concessionaire's support for the idea.

The Perception of Nature as Under Threat

The realisation that people could exert a profound effect on their surroundings and even degrade nature in a conspicuous way can be traced back to the Enlightenment. Yet widespread belief that Americans (and citizens of industrialised nations as a whole) were somehow out of balance with the natural world became particularly acute during the 1960s and 1970s. Prior to that time, environmental concern about the national parks and forests centred on preserving wilderness, loosely defined as an area lacking roads but suitable for 'primitive' recreation in places substantially devoid of changes brought by an industrial society. During the 1930s, some supporters of the national parks quietly rebuked the NPS for excessive road building, while the USFS tried to capitalise on this discontent by administratively designating 'primitive areas', implying that these were more pure than national parks in lacking roads, hotels and tourist amenities. After the Second World War, however, pressure to cut more timber in the national forests prompted wilderness advocates to seek legislation that would make the primitive areas inviolate from both roads and extraction. Success came in 1964 and the enacted bill directed that both the USFS and NPS inventory, recommend which of its lands should be added to a legally recognised 'National Wilderness Preservation System' (Sutter 2002).

Compliance with the Wilderness Act's inventory provisions at Crater Lake National Park took the form of closing roads constructed for administrative use to suppress fires in that eight-tenths of the park located away from Rim Drive and developed areas. Annual park visitation had grown to more than 500,000 by 1966, but most back-country roads were now trails for hikers who never numbered more than 2% of all visitors, even when counting those who traversed a road forming the Pacific Crest Trail (a walking route intended to link the Mexican and Canadian borders) through the park. The NPS formulated a recommendation to Congress involving some 80% of the park to Congress in 1974, but lawmakers took no action on it over the next 30 years.

Strangely enough, the wilderness recommendation did not include Crater Lake, mainly because tour boats (the act generally disqualified areas where there was pre-existing motorised access) still plied its waters. This omission generated little interest among journalists or park supporters and passed almost without comment. Since private vessels had long been banned, few people gave much thought to the motorised tour boats and storage facilities on Wizard Island that had existed in one form or another since 1907. The idea that Crater Lake's clarity might be declining was, however, a different matter entirely. It had set a world record in this regard in 1969, when academic researchers worked to establish a series of baseline measurements. When one of the scientists returned in 1978 to find an apparent decrease in clarity, his first thought was that the culprit could be nutrient enrichment, since Crater Lake's blue colour derives largely from the combination of depth and the scarcity of nutrients such as nitrate. A similar decline in other places such as Lake Tahoe had been traced to human causes such as leaking septic systems, so public concern prompted the NPS to establish a monitoring programme at Crater Lake, beginning in 1982. Over the next decade, Crater Lake set another world record for lake clarity, yet researchers found that it could fluctuate month to month according to a number of factors (Harmon 2002).

Researchers have long known that lakes are good indicators of changes within their drainage basins, whether caused by humans or not. Visitors could not discern whether Crater Lake's clarity was declining (it is measured from the lake surface with an instrument called a Secchi disc), but the suggestion that human activity might compromise one of the timeless qualities associated with sublimity made it into something of a crucible for the way in which the NPS met threats to the central features of the parks. When drilling for geothermal energy commenced outside park boundaries in 1986, conservation groups and several scientists expressed the fear that such activity might disrupt lake processes, given its volcanic origins. To stop the drilling, researchers needed to show active vents emitting significantly warmer water than the ambient temperatures at the bottom of Crater Lake. Given this perceived threat, the NPS found additional funds while its monitoring programme provided the infrastructure for a manned submersible to explore the bottom for the first time ever in 1988. Researchers made a return visit in 1989 to great media fanfare in Oregon, eventually gathering enough evidence to have the drilling halted (Harmon 2002).

Annual visitation actually fell below 500,000 during the two summers in which the submersible explored the lake's depths, to a point where Lassen Volcanic National Park in northern California (which traditionally lagged behind its northern neighbour) became a bigger draw than Crater Lake for the first time. The latter, however, received substantial funding increases to support additional staff in natural resources management which allowed it to build the leading lake monitoring programme among all NPS units. This came concurrently with more agency support for 'resource management' as a distinct endeavour from visitor services (such as interpretation) or facility maintenance, signalling that the NPS no longer solely depended on visitation as an index to its budget and staffing. This change reflected how sublime nature was now conceptualised by many in the NPS and some key supporters outside the agency as a patient, one whose health had to be actively managed through 'vital signs' in order to prevent irreversible decline.

Conclusion

While the capability to manage natural resources throughout the park seemingly improved during the 1990s, the NPS also received increased funding to maintain or even increase its infrastructure that supported visitor services at Crater Lake. Some of it was justified to rehabilitate 'cultural resources' such as the main hotel and other structures that dated from the period when rustic architecture dominated park facilities, even though annual visits to Crater Lake over the decade remained flat at more or less a half-million. The park's place within the wider NPS galaxy as a small national park (one firmly lodged at mid-table with regard to its budget compared to all units) remained unchanged since the 1960s, but overall funding for the agency increased as the American economy boomed during the 1990s. This put the NPS in a better position to fund park operations at Crater Lake at unprecedented levels, although there is evidence of budget contraction since 2000.

Current NPS staffing and budget at Crater Lake largely reflect a layering process that began with how the lake should fit a cultural construction called the sublime. It maintained an iconic stature as a widely recognised focal point in any tour of the Pacific Northwest, yet the indicators of visitation and park budget showed an opposite trend over the past half-century with respect to Crater Lake's place among units administered by the NPS. This 'scenic hierarchy' proved to be somewhat superficial or even irrelevant when the prospect of declining clarity or potential impact of geothermal drilling surfaced as perceived threats during the 1980s. Among other lakes in the western United States, only Lake Tahoe and Mono Lake have generated similar support against environmental degradation, but then only when documented changes had been well underway. The concern at Lake Tahoe was eutrophication, or the possible changes triggered by more nutrient input, whereas changes associated with upstream water withdrawals through time dramatically lowered the level of Mono Lake. Just the hypothetical possibility of detrimental change at Crater Lake, by contrast, brought an immediate response of monitoring and basic research – perhaps because it symbolises the eternal quality of beauty to so many visitors. At the same time, it is the world's most studied caldera lake, so 'iconic stature' in this case may indeed be a projection of what people continue to see as both aesthetically pleasing and intellectually engaging.

References

Cosgrove, D.E. (1998) *Social Formation and Symbolic Landscape*. Madison, WI: University of Wisconsin Press.

Diller, J.S. (1902) *The Geology of Crater Lake National Park, Oregon*. USGS Professional Paper No. 3, Washington, DC: Government Printing Office.

Dutton, C.E. (1886) Crater Lake, Oregon: A proposed national reservation. *Science* 7, 179–82, 26 February.

Harmon, R. (2002) *Crater Lake National Park: A History*. Corvallis, DR: Oregon State University Press.

Havlick, D.G. (2002) Behind the wheel: A look back at public land roads. *Forest History Today* (Spring), 11.

Inquirer (1865) The Sunken Lake. Philadelphia, 17 August .

Mark, S.R. (2002) A study in appreciation of nature: John C. Merriam and the educational purpose of Crater Lake National Park. *Oregon Historical Quarterly* 103 (1) (Spring), 98–123.

Miller, J. (1904) The Sea of Silence. *Sunset* 13 (5) (September), 404.

Pomeroy, E. (1990) *In Search of the Golden West: The Tourist in Western America*. Lincoln, NE: University of Nebraska Press.

Ross, S. (1998) *What Gardens Mean*. Chicago: University of Chicago Press.

Runte, A. (1990) *Yosemite: The Embattled Wilderness*. Lincoln, NE: University of Nebraska Press.

Smith, R.A. (1959) Porter pushes Crater Lake chairlift idea. *Mail Tribune* (Medford, Oregon), 6 July.

Smith, R.A. (1960) Ullman opposes Porter's Crater Lake chairlift idea. *Mail Tribune* (Medford, Oregon), 11 February.

Steel, W.G. (1886) Crater Lake and how to see it. *The West Shore* 12 (3) (March), 104–6.

Steel, W.G. (1888a) In the Josephine Caves. *Oregonian* (Portland), 10 September.

Steel, W.G. (1888b) Visit to Crater Lake. *Oregonian* (Portland), 11 September.

Sutter, P.S. (2002) *Driven Wild: How the Fight against Automobiles Launched the Modern Wilderness Movement*. Seattle: University of Washington Press.

Tidings (1877) (Ashland, Oregon), 21 September.

United States Department of the Interior – National Park Service (USDI-NPS) (1997) Draft Visitor Services Plan/Environmental Impact Statement, Crater Lake National Park, Oregon. Washington, DC: United States Department of the Interior – National Park Service.

Chapter 4
The Changing Historical Dimensions of Lake Tourism at Savonlinna: Savonlinna – The Pearl of Lake Saimaa. Lake Representations in the Tourist Marketing of Savonlinna

KATI PITKÄNEN AND MIA VEPSÄLÄINEN

The lake landscape in Figure 4.1 is a typical way of representing Savonlinna in the tourist brochures. Physically it encompasses natural as well as artificial elements. The lake landscape is primarily a natural environment, consisting of historically formed lakes, islands and shorelines, but it can also include various artificial and man-made elements, such as steamboats, cargo vessels, bridges, summerhouses and industrial plants. Besides the physical appearance, the lake landscape also has a more abstract side. Most of the characteristics we associate with lake scenery can not be identified by any physical trait. Those include for example memories or dreams – thus, meanings or qualities related to the place and its past, present and future. These symbolic meanings, evoked by both the natural environment and man-made elements, are always historical and cultural. They are meanings that people give to places (Short 1996; Äikäs 2001; see also Carr, this volume; Mark, Chapter 3 this volume).

The meanings attached to different types of landscapes in different times and among different cultures are diverse. According to Meining (1979: 164) *'every mature nation has its symbolic landscapes. They are part of the iconography of nationhood, part of the shared ideas and memories and feelings which bind a people together'*. Many studies of the British, North American and Australian landscape have stated that especially countryside or wilderness landscapes are often established as symbolic or ideal (e.g. Osborne 1988; Short 1991; Bunce 1994; Matless 1998). For example, Osborne (1988) emphasised the importance of northern nature for the formation of the cultural identity of Canada. He found that the explicitly

Figure 4.1 Picture from a tourist brochure entitled 'Savonlinna à la Carte': St Olof's Castle, a steamboat and lake landscape

northern qualities of nature such as rocky landscape, lakes, forests and white winter became national symbols. Nature has also attained symbolic meanings in Nordic countries that have a very similar natural environment to Canada, with the Finnish lakes being considered a symbol of Finnish nationalism.

National iconography, however, is not timeless, as for example Matless (1998) observed in his study of the national symbolism of the English landscape. According to him, the British idealisation of the countryside attained remarkably many modern features during the 20th century. This has resulted from the ideal of modern planning that has enforced the preservation of symbolic landscapes. Similarly, modern elements have influenced the appreciation of Finnish lake landscape.

National landscapes and other symbolic landscapes are communicated to people mainly through different media. The abstract meanings of landscapes are reproduced through numerous visual and textual landscape representations (Meining 1979). In this chapter we will look at the different meanings the Finnish lake landscape has conveyed in different times. We will approach the meanings by analysing the historical lake landscape representations of the tourist marketing material of Savonlinna.

Landscape Representations in Brochures

The contents of this chapter should be seen within the context of the new cultural geography that emerged in the discipline of human geography with the 'cultural turn' during the 1970s and 1980s. It therefore focuses especially on the social interpretation of landscape and landscape representations. However, the concept of landscape has been contested in the new cultural geography. Whereas the traditional landscape geography aimed at classifying different types of landscapes, the new landscape geography has paid attention to the ideological, symbolic and power-laid structures behind seeing the land as a landscape. Hence, instead of being a tool for making spatial distinctions, landscape is defined as a socially and culturally influenced way of perceiving the world (e.g. see Mitchell 2000).

With the new landscape geography, visual and written descriptions of landscapes have been highlighted as an important aspect of landscape research (Mitchell 2000: 60–3). In the landscape experience, the representations are considered equal to the actual sensed environment – for example, the pioneering research on landscape presentation contained in Cosgrove and Daniels (1988) which discusses the symbolic meanings of landscapes represented in art, maps and promotional material. The representations in advertisements and brochures have also been widely analysed by, for example, Burgess (1982), Gold and Gold (1990), Gold (1994), Ward (1994, 1998) and Hopkins (1998a).

Landscape representations in brochures and other tourist marketing material are tightly intertwined with place promotion. Ward and Gold (1994) define place promotion as the intentional use of publicity and marketing to communicate certain images of places to a particular target group. The overall aim is to affect the behaviour of the target group and make the destination tempting for entrepreneurs, tourists and local residents (Ashworth & Voogd 1990; Holcomb 1993; Paddison 1993; Hall 1998).

Even if the marketing material is planned to promote the particular interests of the advertisers, it cannot be interpreted without taking the social and historical circumstances in which they were produced into account. Thus, the preferences of the advertisers are often related to certain times and spaces (Burgess & Wood 1988; Sadler 1993; Ward 1994). This means that the representations in the promotional material get selected according to the socially evident power relations and usually tend to reinforce them (Morgan & Pritchard 1998; Short & Yeong-Hyun 1998). In the case of the lake landscape the images promoted in the brochures may be interpreted as reflecting the general and historical preferences of a desired lake environment (Lewis 1988; Hovi-Wasastjerna 1996). Hence, the pictures and written descriptions of lake landscape in the brochures can be interpreted as cultural representations that convey the meanings attached to the landscape.

In the following sections, after a short introduction to Savonlinna as a tourist destination, we will look at the lake representations of Savonlinna's tourist advertising since the beginning of the 20th century. The analysis concentrates on the representations of Savonlinna and the surrounding lake Saimaa communicated and promoted in travel films and brochures. The analysis indicates that the different types of lake representations and the meanings attached to them can be divided into four distinctive categories in which the lake landscape has been used to create or convey different images:

- Finnishness (since the end of 19th century)
- nostalgia (since 1960)
- holiday experience (since 1970)
- nature (since 1980)

Savonlinna as a Tourist Destination

Savonlinna is situated in the middle of Lake Saimaa, the largest lake district of Finland. Saimaa is a network of lakes and rivers that connects the whole of eastern Finland. Through the Saimaa canal it also has a connection to the Baltic Sea. Therefore, Saimaa has historically been one of the most important inland traffic lanes in Finland, making Savonlinna an important port. Today, the town is primarily known for its medieval St Olof's Castle and international opera festival. Historically, however, the surrounding lake scenery has been a central element in the construction of Savonlinna's image. The lake labyrinth of Saimaa with its (steam)boat traffic and romantic landscape has attracted tourists for over a hundred years.

The central location, the castle and the beautiful surroundings, especially the picturesque ridge and lake landscape in Punkaharju, made Savonlinna one of the most popular sights of the late 19th century in Finland. At first, the town was known as a place for quick sightseeing tours, but later, after the founding of an opera festival and a spa in the beginning of the 20th century, the character of the tours changed. The contemporaneous appearance of passenger steamboats on Lake Saimaa and the improvements in the infrastructure of overland transport led to a dramatic increase in the amount of passengers. From the turn of the 20th century until the Second World War, both Savonlinna and the surrounding Lake Saimaa had a central role in tourism at national level (Vehviläinen 1978).

Passenger traffic recovered fast from the depression following the two world wars. For example, the rate of visitors to Olavinlinna reached pre-war levels by the late 1950s. However, it was not easy for Savonlinna to regain its position as one of the leading tourist destinations in Finland. Competition for both foreign tourists and urbanised Finnish holidaymakers had got stronger. New areas and destinations, such as Lapland, appeared on the tourist map of Finland and neither the beautiful lake landscape nor the romantic walls of Olavinlinna were enough to attract modern tourists seeking amenities and pleasure. In order to overcome the challenges, new tourist marketing organisations were set up inside and between the municipalities of the Savonlinna region. The opera festival was revived in 1967 and actions were taken to build a new infrastructure and renovate the old one. This included restoring the castle for its 500th anniversary in 1975 (Vehviläinen 1978).

Today, tourism plays an important role in the economic and cultural life of Savonlinna. Tourism's marketing of Savonlinna is still centred on Olavinlinna Castle, the annual opera festival and the scenery. However, the new marketing activities, based on worldwide trends in tourism, emphasise the role of activity in the creation of the tourist experience.

Finnishness Since the End of 19th Century

The oldest tourism marketing material of Savonlinna and the surrounding lake district dates back to the early 1900s. The first marketing pictures presented in the brochures can be reduced to three distinctive types of representation: the monumental St Olof's Castle rising from a lake, a horizontal lake landscape pictured from the Punkaharju ridge and a panoramic lake landscape seen downwards from an outlook tower (see Table 4.1). Written descriptions of the lakes also emphasised the aesthetics of the landscape:

> Punkaharju is a high narrow ridge, four miles long but so narrow that from the drive which runs along the top of the ridge, a stone can be thrown in to the clear waters which lap its steep pine covered shores. Lovely views of the islands which thickly dot the surrounding lakes are to be seen on both sides of the ridge. (Finland: The land of a thousand lakes 1909)

The earliest lake representations were intertwined with the image of the lakes as a national landscape. Finland gained independence in 1917, although a nationalist ideology had become well developed by the second half of the 19th century. Nationalism was systematically promoted by the leading opinion formers of Finland's political as well as cultural life. Because the national history of Swedish then Russian occupation could hardly serve as the basis for nation-building, the Finnish national movement resorted to the use of the nation's exceptional qualities of the natural environment (Häyrynen 1994; Paasi 1996; Eskola 1997). The lake

Table 4.1 Representations of Finnishness in the advertising material of Savonlinna

	1900	1930	1950–1970	1980	1990
Finnishness	The monumental St Olof's Castle rising from a lake A horizontal lake landscape pictured from the Punkaharju ridge A panoramic lake landscape seen downwards from an outlook tower	Lake landscape covered with ships, bridges, shoreline road and railway networks, rafts of logs and industrial plants Aerial photographs of the lake landscape Pictures of the lake landscape taken from a moving vehicle	Colour prints Lake landscape in the winter	Lake landscape in the night-time	Pictures of lake landscape with a deep saturation of blue and green

landscape was chosen as the national landscape at the beginning of the century. It fulfilled the aesthetic preferences of the time and was also a geographically important glacial monument, as well as being culturally Finnish in relation to the surrounding Swedish- and Russian-speaking areas (Häyrynen 1994; Vasenius 1996; Eskola 1997). The physical qualities of the natural surroundings adopted abstract meanings of national superiority. The lake landscape also reflected the features of the nation and its people – freedom, uniqueness and perseverance.

The romantic nationalism also gave an impetus to domestic tourism and travelling became more common among the upper classes at the end of the 19th century. The primary ambition of tourism was to enhance knowledge of the fatherland and travelling was tightly intertwined with the romantic nationalist image of Finland. The status of Lake Saimaa as a national landscape made Savonlinna and the surrounding lake region one of the most important domestic tourist destinations (Eskola 1997; Hirn & Markkanen 1987). The lakes were also a physical prerequisite for tourism. Savonlinna had access to the other towns in the area along the waterways and the Saimaa canal all the way to the Baltic Sea. Boat traffic flourished and the harbour filled with white steamboats.

The variety of lake representations grew in the 1930s. The new motifs of the time included pictures of the lake landscape covered with ships, bridges and shoreline road and railway networks, as well as rafts of logs and industrial plants (see Table 4.1). These were signs of the changes the physical environment went through during the modernisation process of Finland. The forest industry especially became a powerful symbol of national self-sufficiency and welfare. Because lakes offered an important traffic lane and a resource for the forest industry, the signs of industrialisation became visible in the lake landscape and thereby in lake representations. For the forest industry, lakes offered an important traffic lane and a resource, and thereby became visible in lake representations. On a symbolic level, the signs of modernisation in brochures reflected collective national pride and were, for example, compared to the national romantic symbols: 'The Castle is a remarkable feat, as is the mighty railway bridge, under which the ship will now be conducted . . . ' (Saimaan alue 1928).

The ethos of modernisation, however, was not emphasised for long and it began to fade away during the 1950s. The more individual and family-centred values emerged in brochures and gradually replaced the collective national meanings of the representations. In contrast, the national romantic lake landscape has retained an important role in the representation of Finnish lake landscape until the present day. New photography techniques, such as aerial photography in the 1930s and colour prints in the 1950s, were adopted to illustrate the labyrinth of lakes and islands (see Table 4.1). The content of the representations and the symbolic meanings of the lake landscape, however, have remained essentially the same.

Nostalgia (since 1960)

The rise of individual values became visible in the 1960s. Brochures were filled with pictures of holidaymakers and elements important for relaxing. Typical new motifs included pictures of lakeside summerhouses and saunas, as well as families rowing and angling on the lake. Similarly, representations of the traditional and peaceful countryside, often seen from the deck of a steamboat, were common (see Table 4.2). Also, brochures used local dialects in their information to create a feeling of familiarity.

The nostalgic meanings attached to the lake representations resulted from the rapidly occurring structural transition in Finland during the 1960s and 1970s. The structural changes in agriculture led to strong inland migration from the countryside to urban areas. People who were forced to move to towns returned to the countryside during their holidays, charged with positive images and traditional values, to compensate the stress of urban life. For the holidaymakers, the lake landscape evoked memories and images of childhood and sunny holidays and linked the landscape to a sense of nostalgia (see Hietala 1998; Kolbe 1999; Raivo 1999). Among urbanised people, it became popular to invest in one's own summer cottage by a lake and the amount of cottages increased dramatically (Krohn 1991). The nostalgic escape to the countryside made the lakeside summer cottage and sauna a symbol of the ideal Finnish holiday.

However, the significance of steamboats changed. Their importance as a means of transport decreased as overland transport, especially in the amount of private cars, increased. Instead, steamboat cruises became

Table 4.2 Representations of nostalgia in the advertising material of Savonlinna

1960–90
Lakeside summer cottages/houses and saunas
Families rowing and angling on the lake
Steamboat cruises
Idyllic countryside: lakeside farmhouses and fields (farm animals, traditional farmwork)
Local atmosphere as an attraction of lake tourism: dialects, local habits, traditions, traditional dishes

popular. For many people, steamboats evoked happy memories of childhood. As in depictions of the romanticised countryside, steamboats played an important part in evoking nostalgic memories for tourists enjoying the lake landscape.

That nostalgia is still visible in advertisements for lakeland holiday attractions. However, people no longer escape town and city life in such numbers, and the latest generation of urban dwellers attach new meanings to the lake landscape.

Holiday Experience (since the 1970s)

In the 1970s, the amount of leisure and tourist facilities represented in brochures increased. The role of the lake landscape was to serve as a setting for various leisure activities practised both on the water and on shore. Typical brochures of the time were illustrated with pictures of waterskiing, seaplanes, boating and various beach activities (see Table 4.3.) Also, the information described the various activities and accommodation possibilities the lakeland had to offer tourists and holidaymakers of different age groups:

> Especially during the summertime Savonlinna is a lively international tourist town, surrounded by lakes, offering a wide range of holiday and recreation facilities, from a first class hotel to an isolated summer villa on the shores of the Saimaa lake system, from a lively holiday village to a quiet camping site on the wilds. (Finland, Savonlinna and surroundings around 1970)

Also, steamboat advertisements adopted new features that highlighted the excitement of the cruises. Besides admiring the scenery, being on a cruise included games on the deck, dining with the captain, petting the ship's parrot, lying on sunloungers and staying the night in a luxurious cabin. Romance was attached to cruises and the representations altogether reflected an atmosphere of a 'loveboat' (the title of a 1970s American television series set on a cruiseship) (see Table 4.3).

These changes in advertising reflect the establishment of a welfare state in Finland in the 1970s. The standard of living increased and people could afford to buy modern conveniences. Tourists also began to demand more of the tourist destination and the ways of travelling changed. Trips abroad became more common and package tours to the sunny beaches of southern Europe became especially popular. Homeland destinations had to face new kinds of competition. This was visible in the tourist marketing of Savonlinna, particularly in the amount of activities and possibilities offered. In addition, ipictures of the lake landscape in some advertisements were very similar to the images of the Mediterranean seaside.

Table 4.3 Representations of the attractions of a 'holiday experience' in the advertising material of Savonlinna

1970	1980	1990
Water-based activities: water-skiing, seaplanes, boating etc. Steamboats as 'loveboats'	Saimaa as a destination for activity and family vacations Theme parks Package tours	Experiences: nature-based activities (canoeing, hunting etc.), winter activities Stories, histories, narratives: e.g. the ghost of St Olof's Castle, the romanticised history of steamboats Adventures: e.g. jet-ski safaris, pirate cruises

The individual and recreational potential of the lake landscape grew even bigger in the 1980s. The trend for images of southern Europe was replaced by advertisements showing almost placeless theme parks and holiday villages (see Table 4.3). However, in comparison with foreign destinations, the attractions of Finland were highlighted as an asset:

> When Finns were recently asked: Where would you spend your summer holiday? Most of them responded: To the attractive Savonlinna region, there is the Opera festival and St Olof's Castle, there are ships and a lake landscape and there is always something going on. That is true also in the winter, when the sun tans you better on the frozen lakes than in the sunny beaches of the south, when the perch bites your jig under the lake's icy cover and when your skis slide smoothly in the snowy lake landscape. (Savonlinna 1983)

By the turn of the 1990s, images of placeless amusements were no longer included in brochures and local attractions became increasingly important. Lake advertisements emphasised experiences, adventures and stories related to the local environment and culture. This resulted in the appearance of pictures of nature-based activities, traditional fishing methods, adventures and the attractions of the lake landscape in winter (see Table 4.3). Also, the written descriptions represented the lake landscape as an experience:

> Have a cruise experience on Lake Saimaa! Magnificent Lake Saimaa with its islands and fish-inhabited waters is a unique experience. Relax on the deck of a steamboat admiring the fabulous scenery. Try your luck on a fishing trip, enjoy the atmospheric night cruises or rent a steamboat for a family celebration, or sail for a berry- or a mushroom-picking trip in the archipelago – just as you want! (Savonlinna cruises 1993)

During the 1990s, the individual values increasingly gained more importance. People started to strive after experiences and actively try to find ways of expressing themselves (Opaschowski 1995; Poon 1996; Mäenpää 1999). Scientifically, the still ongoing process has been interpreted as a cultural transition into a post-modern 'dream society' where consumption is based on images rather than the actual qualities of products (see Jensen 1999; Hopkins 1998b; Jessop 1998). In tourism, the post-modern changes have arguably led to the birth of a new type of tourists. 'The new tourists' are more independent, flexible and experienced than former generations, but they travel to the lakes to experience the aesthetic landscape and excitement of the activities. Besides the aesthetic and exotic experiences and adventures that the lake scenery has to offer, the lakes can also function as an ecological experience of a pure and clean environment (see Poon 1996).

Environmental Consciousness (since the 1980s)

The quality of the lake environment became a popular theme in advertising literature at the beginning of the 1980s. The clean waters and conservation value of the lakes were distinctly emphasised: 'back to nature ... the most beautiful and cleanest countryside in Europe ... Saimaa Lakeland.' At that time, the Saimaa seal (*Phoca hispida saimensis*), which today is a well-known symbol of the lake district, also appeared in the advertisements (see Table 4.4).

The appearance of ecological factors in advertising literature is a result of the emergence of appreciation of environmental values at a global level. In the West, the increase in damage to the environment and the growing awareness of the limitations of in natural resources led to criticism of the development ethos in the 1960s. The growth of environmental consciousness led to a breakthrough in sustainable development ideology in the 1980s and awareness of the environment became a part of everyday life in the West. Consequently, environmentalism could be used as an asset in tourism marketing.

Table 4.4 Representations of environmental consciousness in the advertising material of Savonlinna

1980	1990
Saimaa seal and salmon	The aesthetic quality of the environment
Cleanness of the lakes	
	Cleanness
Conservation	

Environmental values were still an important part of the tourist brochures of Savonlinna in the 1990s. Instead of the ecological values such as conservation and a pollution-free environment however, advertisements emphasised the beauty of a clean and authentic lake landscape. During the 1990s, the lake landscape has become a tourist destination and a product for the new environmentally orientated tourists seeking authentic and individual experiences. Also, the tourist destinations are required to offer more personal services and offer authentic experiences in an environmentally sustainable manner (Poon 1996).

Discussion

In this chapter we have approached lake landscape representations in the historical advertising material of Savonlinna. We have shown how the different types of representations and the meanings attached to them can be divided into four categories. Related to time, the categories form a historical continuum overlapping each other. The contemporary representations encompass meanings, some of which are new and some of which have accumulated during the past century.

The categories of lake landscape representation are narratives that reflect the changing historical and cultural reality. Even though the analysed material consists of tourist brochures, the meanings attached to the lake landscape have a wider cultural origin. The lake landscape has adopted different meanings according to what has been regarded as desirable or wanted in relation to both social development and individual values. Hence, the categories relate the importance of the lakes and lake landscape to Finns and Finnishness.

Besides the physical features, the lake landscape also has an abstract side. People give meanings to their surroundings and the places they visit or live in. Similarly, besides the actual physical environment, the symbolic meanings are essential to the tourist marketing of places. However, the natural environment and the cultural meanings attached to it are relatively unchangeable. For example, the Finnish lake landscape was formed during the last Ice Age and, as we have shown, some of the meanings attached to it are over a hundred years old. Therefore, it can be argued that the lake landscape is static to the needs of swiftly changing tourism marketing trends (see Paddison 1993; Ashworth & Voogd 1990).

At the beginning of the 21st century, the tourism marketing of places has become more diverse, but at the same time more universal. Despite the need to stand out from the masses, the place-specific promotion material rarely conveys any special or personal features. The local history, landscape, special sights and activities are represented in a manner that is very similar to almost anywhere in the world. The outcome is brochures, films and marketing strategies that offer anything to anyone and create images that can be located anywhere (Ashworth & Voogd 1990; Philo & Kearns 1993; Barke & Harrop 1994; Short & Yeong-Hyun 1998). The 21st century promotional material of Savonlinna also merges the past and present in an almost absurd and somewhat bizarre manner. In addition to the new experiences, the old approved images are used to attract all sectors of the tourist market.

Nevertheless, it be successful, the image of a tourist site being promoted must be based on physical reality, as well as on the prevailing cultural and social interpretations of it. In the case of the marketing of tourism in Savonlinna, it would be important to consider what the images are that the lake landscape can – and even more importantly – cannot attain. More specifically, what does the lake landscape signify to tourists, what are the abstract meanings attached to the physical reality and how sustainable a solution is it to alter the lake environment to correspond to the needs of current trends in tourism and tourism marketing?

References

Äikäs, T A. (2001) Imagosta maisemaan: esimerkkinä Turun ja Oulun kaupunki-imagojen rakentaminen (From images into landscapes: cases of the building of the city images of Oulu and Turku). *Nordia Geographical Publications* 30 (2).

Ashworth, G.J. and Voogd, H. (1990) *Selling the City: Marketing Approaches in Public Sector Urban Planning*. London: Belhaven Press.

Barke, M. and Harrop, K. (1994) Selling the industrial town: Identity, image and illusion. In J.R. Gold and S.V. Ward (eds) *Place Promotion: The Use of Publicity and Marketing to Sell Towns and Regions*. (pp. 93–114). Chichester: John Wiley & Sons.

Bunce, M. (1994) *The Countryside Ideal: Anglo-American Images of Landscape*. Routledge, London.

Burgess, J. (1982) Selling places: Environmental images for the executive. *Regional Studies* 16 (1), 1–17.

Burgess, J. and Wood P. (1988) Decoding docklands. Place advertising and decision-making strategies of the small firm. In J. Eyles and D.M. Smith (eds) *Qualitative Methods in Human Geography* (pp. 94–117). Padstow: Polity Press.

Cosgrove D. and Daniels, S. (eds) (1988) *The Iconography of Landscape: Essays on the Symbolic Representation, Design and Use of Past Environments*. Cambridge: Cambridge University Press.
Eskola, T. (1997) *Teräslintu ja lumpeenkukka: Aulanko-kuvaston muutosten tulkinta* (*Interpretation of the Historical imagery of Hill Aulanko*). Helsinki: Musta taide.
Gold, J.R. (1994) Locating the message: Place promotion as image communication. In J.R. Gold and S.W. Ward (eds) *Place Promotion: The Use of Publicity and Marketing to Sell Towns and Regions* (pp. 19–37). Chichester: John Wiley & Sons.
Gold, J.R. and Gold M.M. (1990) 'A place of delightful prospects': Promotional imaginary and the selling of suburbia. In L. Zonn (ed.) *Place Images in Media: Portrayal, Experience and Meaning* (pp. 159–82). Rowman & Littlefeld.
Hall, T. (1998) Introduction, Part I: Selling the entrepreneurial city. In T. Hall and P. Hubbard (eds) *The Entrepreneurial City: Geographies of Politics, Regime and Representation* (pp. 27–30). Chichester: John Wiley & Sons.
Häyrynen, M. (1994) *National Landscapes and their Making in Finland. Undercurrent*. Available at www.uoregon.edu/~ucurrent/1.4.html. Accessed 15 April 2002.
Hietala, V. (1998) Maalta olet sinä tullut... – Maaseutu, koti ja kansakunnan myytti elävän kuvan kulttuurissa (Countryside, home and the myth of a nation in the Finnish film culture). Wider Screen 1/98. Available at www.film-o-holic.com/widerscreen/1998/1/index.htm.
Hirn, S. and Markkanen, E. (1987) *Tuhansien järvien maa: Suomen matkailun historia*. (*History of Tourism in Finland*) Matkailun edistämiskeskus ja Suomen matkailuliitto. Gummerus, Jyväskylä.
Holcomb, B. (1993) Revisioning place: De- and re-constructing the image of the industrial city. In G. Kearns and C. Philo (eds) *Selling Places: The City as Cultural Capital, Past and Present* (pp. 133–44). Oxford: Pergamon Press.
Hopkins, J. (1998a) Commodifying the countryside: Marketing myths of rurality. In R. Butler, C.M. Hall and J. Jenkins (eds) *Tourism and Recreation in Rural Areas* (pp. 139–56). Chichester: John Wiley & Sons.
Hopkins, J. (1998b) Signs of the post-rural: Marketing myths of a symbolic countryside. *Geografiska Annaler* 80B (2), 65–81.
Hovi-Wasastjerna, P. (1996) *Onnen avain: Mainoselokuva Suomessa 1950-luvulla* (Finnish 1950s commercials). Espoo: Frenckellin kirjapaino Oy.
Jensen, R. (1999) *The Dream Society: How the Coming Shift from Information to Imagination will Transform your Business*. New York: McGraw-Hill.
Jessop, B. (1998) The narrative of enterprise and the enterprise of narrative: Place marketing and the entrepreneurial city. In T. Hall and P. Hubbard (eds) *The Entrepreneurial City: Geographies of Politics, Regime and Representation* (pp. 77–99). Chichester: John Wiley & Sons.
Kolbe, L. (1999) Suomalainen kaupunki (A Finnish city). In M. Löytönen and L. Kolbe (eds) *Suomi – Maa, kansa, kulttuurit* (Finland – nation, people and cultures) (pp. 156–70). Helsinki: Suomalaisen kirjallisuuden seura.
Krohn, A. (1991) *Elämäni lomassa: Suomalaisen loman historiaa* (*History of the Finnish Holiday*). Helsinki: Lomaliitto.
Lewis, M. (1988) Rhetoric of the Western interior: Modes of environmental description in American promotional literature of the nineteenth century. In D. Cosgrove and S. Daniels (eds.) *The Iconography of Landscape: Essays on the Symbolic Representation, Design and Use of Past Environments* (pp. 179–93). Cambridge University Press, Cambridge.
Mäenpää, P. (1999) Hyvinvointiyhteiskunnasta kulutuskulttuuriin (From welfare society into consumer society). In M. Löytönen and L. Kolbe (eds). *Suomi – Maa, kansa, kulttuurit* (Finland – nation, people and cultures) (p. 174). Helsinki: Suomalaisen kirjallisuuden seura.

Matless, D. (1998) *Landscape and Englishness*. London: Reaction Books.
Meining, D.W. (1979) Symbolic landscapes. Some idealizations of American communities. In D.W. Meining (ed.) *The Interpretation of Ordinary Landscapes: Geographical Essays* (pp. 164–92). Oxford: Oxford University Press.
Mitchell, D. (2000) *Cultural Geography: A Critical Introduction*. Oxford: Blackwell.
Morgan, N. and Pritchard, A. (1998) *Tourism Promotion and Power: Creating Images, Creating Identities*. Chichester: John Wiley & Sons.
Opaschowski, H.W. (1995) *Freizeitökonomie Marketing von Erlebniswelten*. Freizeit- und Tourismusstudien. Opladen: Leske & Budrich.
Osborne, B.R. (1988) The iconography of nationhood in Canadian art. In D. Cosgrove and S. Daniels (eds) *The Iconography of Landscape: Essays on the Symbolic Representation, Design and Use of Past Environments* (pp. 162–78). Cambridge: Cambridge University Press.
Paasi, A. (1996) *Territories, Boundaries and Consciousness. The Changing Geographies of the Finnish–Russian Border*. Chichester: John Wiley & Sons.
Paddison, R. (1993) City marketing, image reconstruction and urban regeneration. *Urban Studies* 30 (2), 339–59.
Philo, C. and Kearns G. (1993) Culture, history, capital: A critical introduction to the selling of places. In G. Kearns and C. Philo (eds) *Selling Places: The City as Cultural Capital, Past and Present* (pp. 1–32). Oxford: Pergamon Press.
Poon, A. (1996) *Tourism, Technology and Competitive Strategies*. Wallingford: CAB International.
Raivo, P.J. (1999) Maisema ja mielikuvat (Landscape and images). In M. Löytönen and L. Kolbe (eds) *Suomi – Maa, kansa, kulttuurit (Finland – nation, people and cultures)* (pp. 70–87). Helsinki: Suomalaisen kirjallisuuden seura.
Sadler, D. (1993) Place-marketing, competitive places and the construction of hegemony in Britain in the 1980s. In G. Kearns and C. Philo (eds) *Selling Places: The City as Cultural Capital, Past and Present* (pp. 175–92). Oxford: Pergamon Press.
Short, J.R. (1991) *Imagined Country: Society, Culture and Environment*. London: Routledge.
Short, J.R. (1996) *The Urban Order: An Introduction to Cities, Culture and Power*. Oxford: Blackwell.
Short, J.R. and Yeong-Hyun, K. (1998) Urban crises/urban representations: Selling the city in difficult times. In T. Hall and P. Hubbard (eds) *The Entrepreneurial City: Geographies of Politics, Regime and Representation* (pp. 55–75). Chichester: John Wiley & Sons.
Vasenius, J. (1996) Suomi-neidon kertomaa – suomalaisen identiteetin syntyminen ja määrittely kansallisten symbolien kautta ensimmäisellä sortokaudella (National symbolism and defining Finnish identity during the first period of oppression 1899–1905). In T. Soikkonen and V. Vares (eds) *Kuva ja historia (Image and History)*, (pp.107–44). Turun Historiallinen Arkisto 50, Turku.
Vehviläinen, O. (1978) *Savonlinnan kaupungin historia III (History of Savonlinna)*. Savonlinnan kaupunki.
Ward, S.V. (1994) Time and place: Key themes in place promotion in the USA, Canada and Britain since 1870. In J.R. Gold and S.W. Ward (eds) *Place Promotion: The Use of Publicity and Marketing to Sell Towns and Regions* (pp. 53–74). Chichester: John Wiley & Sons.
Ward, S.V. (1998) *Selling Places: The Marketing and Promotion of Towns and Cities 1850–2000*. New York: Routledge.
Ward S.V. and Gold, J.R. (1994) Introduction. In J.R. Gold and S.W. Ward (eds) *Place Promotion: The Use of Publicity and Marketing to Sell Towns and Regions* (pp. 1–17). Chichester: John Wiley & Sons.

Brochures

Finland: The land of a thousand lakes (1909) A New Holiday Resort. The Tourist Society of Finland.
Finland, Savonlinna and surroundings (~1970). Saimaan matkailu r.y.
Saimaan alue (1928) (Lake Saimaa region). Suomen matkailijayhdistys, Kuopion osasto.
Savonlinna (1983) Savonlinnan matkailupalvelu r.y.
Savonlinna cruises (1993). Risteilyalukset (cruisers). Saimaan matkailupalvelu.

Chapter 5
Lakes, Myths and Legends: The Relationship Between Tourism and Cultural Values for Water in Aotearoa/New Zealand

ANNA CARR

Indigenous people are often characterised as being inseparable from the landscapes they inhabit through the use of resources or symbolic and historical associations with natural environments. The physical presence of indigenous people is obvious to visitors – for example, traditional homes, monuments, archaeological sites or villages. The intangible aspects of culturally significant landscape features are often elusive, but visitors may become aware of them through promotional material; the designation of indigenous place names; and through interpretation which usually provides information of local history, the use of natural resources and local myths or legends (Upitis 1989; Keelan 1996; Hinch 1998; Avery 1999; Brown 1999; Kearsley et al. 1999; Pfister 2000; Shackley 2001). This chapter introduces the importance of the natural environment to the indigenous people of Aotearoa/New Zealand, the Maori, and gives examples to illustrate how visitors may be informed of the cultural values for lakes in New Zealand.

Maori and Lakes

The New Zealand Maori have a close relationship with the natural environment. New Zealand has extensive natural areas offering a wide range of recreational opportunities for visitors. Many of the natural areas have significant cultural values to *tangata whenua* (original inhabitants of the land, indigenous peoples, Maori) and *pakeha* (non-Maori, European) that can be harmed through the inappropriate activities of tourists (O'Regan 1990; Keelan 1996). Mountains, lakes, rivers, coastal and forest areas contain resources and features that are considered to be *taonga* (treasured) by Maori (Walker 1992; Sinclair 1992a, 1992b). Specific landscape features and natural resources, including those associated with water, are traditionally viewed as having a *mauri* or life force. Traditional villages were often sited near lakes which were an important source of *mahinga kai* (food, also places where food is stored, such a freshwater fish and eel). *Tapu* (sacred) sites, including *urupa* (burial grounds), were often located on the shores or islands of rivers and lakes. Lakes have also been associated with traditional myths, tribal legends and could be the dwelling places of supernatural beings such as *taniwha* (monster, powerful person).

Managing the Cultural Landscape

The Conservation Act (1987) and Resource Management Act (1991) guide the Crown agencies and government departments, such as the Ministry for the Environment and Department of Conservation (DOC), regional or district councils in the management of New Zealand's natural and urban environments. The acts purport to conserve the natural resources of New Zealand by controlling the scale and type of industry, and other activities – for example, tourism or recreation – which may occur in various environments, including lakes. Both Acts reflect the principles of the Treaty of Waitangi, the founding constitutional document of New Zealand. Article Two of the Treaty of Waitangi intended to guarantee Maori their exclusive rights and interests to collective land, resources and *taonga* (things of value, treasured) (Sinclair 1992a, 1992b). From the mid-1970s an intense revival of pride in Maori culture resulted in increased Maori pressure for government recognition of Maori traditions, including the *kaitiaki* (traditional guardianship) of natural resources (Walker, 1992). In 1975 the Waitangi Tribunal was established to settle Maori land and resource claims surrounding Treaty grievances. The past two decades has seen an increased awareness of treaty obligations by the New Zealand government with consequent legislation reflecting the 'spirit' of the Treaty of Waitangi. Treaty settlements increasingly recognise that local *iwi* have *kaitakitanga* of natural resources, lakes, rivers, ancestral sites and other *taonga*.

Cooperative conservation management encouraging consultation with *iwi* (tribal groups) and acknowledgement of *kaitiakitanga* (traditional guardianship) are important principles of natural resource management in New Zealand. One aspect of Treaty settlements that affects lakes usage has been the designation of *nohoanga* or *ukaipo* (camp sites) on Crown land that enable Maori belonging to specific *iwi* to enjoy the traditional practices of gathering food and natural resources. *Nohoanga* are usually up to one hectare in size and are often located by lakes – for instance, Mavora Lakes in Central Southland; Lake Kaniere in Westland; Lake Pukaki in South Canterbury and Lake Waikaremoana in the North Island.

Previous studies have noted that Maori are poorly represented as participants in outdoor recreation in natural areas (Lomax 1988; Matunga 1995a, 1995b; Devlin and Booth 1998). Maori outdoor recreation was viewed as a means of

> linking people and place, or *tangata whenua* with their *turangawaewae* (standing place from where one gains the authority to belong), exploring the natural environments and cultural traditions of their *tupuna* [ancestors; grandparent (pl: tipuna)], reinforcing basic values of Maori culture and instilling a sense of cultural pride. (Matunga 1995a: 18)

The availability of such sites for traditional purposes is a positive initiative as the sites encourage Maori *iwi* members to participate in recreation activities such as fishing, boating and camping in national park and outdoors settings.

Despite the positive social and cultural benefits of *nohoanga*, there has been public opposition. The designation of *nohoanga* has been perceived as racist when the land-use has been exclusively for Maori (*Otago Daily Times* 1998; *Southland Times* 2000). The public can usually camp at nearby sites, public access is maintained to adjacent rivers and lakes and there are time limits as to when the camping areas are set aside as *nohoanga* for exclusive use by *iwi* members. The *nohoanga* are often administered by *rananga* or tribal councils, occasionally utilising European management techniques – for example booking systems to manage usage of such sites. The *rananga* also inform the general public of the *nohoanga* site designation through advertisements within daily newspapers.

Tourism and cultural values

The economic and recreational significance of New Zealand's natural landscape has been instrumental in the development of what is now New Zealand's primary generator of international monetary exchange, with 2.3 million international visitors spending NZ$7.4 billion in the year ending October 2004 (Tourism New Zealand 2004). The tourism industry has a history of promotional imagery, placing Maori alongside the picturesque landscape to add an 'exotic cultural flavour' to potential visitor experiences (Ateljevic & Doorne 2002). Indeed, 'Maori carvings and/or Maori maidens dressed in traditional costume were integrated as essential elements of the "scenic splendour"' (Ateljevic & Doorne 2002: 655). Integrating visual elements of the culture within tourism promotion has continued with textual allusion such as references to Maori mythology and legends within scenic landscapes.

Maori tourism has been defined as 'tourism products that utilise cultural, historical, heritage or natural resources that are uniquely Maori with substantial Maori ownership and control of the business' (Ingram 1997: n.p). Maori tourism operations have traditionally provided cultural performances within hotel or *marae* (community buildings and surrounding area associated with *iwi*) settings, with increasing diversification of the Maori tourism product to include nature-based tourism attractions such as ecotourism in natural areas with cultural significance (Aotearoa Maori Tourism Federation 1994). Increasingly, there are a number of Maori and non-Maori tourism operators and organisations that present the cultural heritage of Maori to visitors, adding a unique cultural dimension to visitor experiences of New Zealand lake environments. Additionally, Maori-owned tourism operations in lake environments can be of economic and social benefit to local *iwi*, connecting owner/operators and staff to their traditional landscapes.

Tourism activities and attractions are often based around lakes that have cultural significance to Maori and attempts have been made to acknowledge the cultural values at a number of lake destinations. Visitors and tourism operations may impact upon Maori values in both positive and negative ways. Economic and social benefits could result if local Maori were involved with the management of commercial visitors operations to significant sites associated with traditional myths, *mahinga kai*, legends or history. However, visitation to sensitive areas, including archaeological sites, *pa* (occupation site or fortified village), battle sites and *waahi tapu* (sacred places), could not only result in negative impacts, but a weakening of the *mana* (power, prestige, authority) associated with such sites. Similarly, local *iwi* who are not involved with resource management are potentially faced with neither having the opportunity to contribute to sustainable tourism development, nor experiencing the financial benefits of any visitation.

Interpreting Lakes

With the Resource Management Act (1991), government agencies are required to consult with local *iwi* regarding the management of culturally significant areas (Department of Conservation 2001). Consultation relevant to lakes tourism has concentrated on public access to and use of, lakes and resources at popular visitor destinations. *Iwi* members may also be consulted about the preparation of visitor information and interpretation that can inform the general public of the cultural values of the area. Cultural interpretation is increasing at a number of significant sites (including lakes) in New Zealand for the specific purpose of educating the public and raising visitor awareness of cultural values in each area (Keelan 1996; MacLennan 2000; Department of Conservation (DOC) 2002b). The cultural values of lakes can be communicated in non-commercial ways – for example, with interpretation panels at picnic and camping areas, road ends, car parks and boat launch ramps. Without such interpretation visitors may be unaware of the significant community and Maori values for the environment, resulting in inappropriate visitor behaviour that causes negative impacts and may harm the *mana* of the area, offending local community members.

Interpretation of tourism activities provided by local *iwi* offering cultural perspectives of the natural world could appeal to a number of visitors seeking a nature-based experience that is both environmentally and culturally sustainable. Occasionally, non-Maori operators seek to offer cultural interpretation within experiences that may have little outward relationship to the culture. For example, the managers of Kawerau Jet in Queenstown have displayed panels narrating the Ngai Tahu legends for Lake Wakatipu on the walls of the business premises.

Mythology complements a range of tourist activities, adding a Maori dimension to the product. Increasingly, companies offering mythology as part of the tourism product are owned by local *iwi* that they have direct control over the final product. The presentation of cultural mythology that is meaningful to *tangata whenua* can be important for strengthening their cultural identity. The retelling of a myth to tourists provides Maori with an opportunity to inform the visitor of their historic ties with culturally significant locations. For contemporary Maori, mythology relates the deeds of their ancestors and explains the traditional view of the formation of the natural world. The association of a myth with a geographical location identifies the spiritual or ancestral significance of a place to the Maori people (Walker 1992). The retelling of a myth provides Maori with an opportunity to convey their *turangawaewae* (attachment to land, sense of place) by informing visitors of their historic, ancestral and spiritual ties with culturally significant areas. Ideally, any commercial use of such information would be managed through the selection of approved material that is communicated with the consent of the traditional owners (Walker 1992, 1996; Keelan 1996; Kearsley *et al.* 1999; Pfister 2000; Staiff *et al.* 2002).

Most lakes destinations in New Zealand provide few insights into cultural values. This could be for a number of reasons. Perhaps local *iwi* do not want to have their values displayed to the visiting public; local marketing and interpretation material may be in a development phase requiring lengthy consultation procedures with *iwi*; there may be a perceived lack of public interest in the material; or such information could have a secondary role – providing the 'exotic' touch to visitor experiences. The following case studies of lake destinations have been selected to illustrate further the *iwi* relationships with lakes and the scale of cultural interpretation that may occur.

Lake Taupo and the Rotorua Lakes

New Zealand's largest lake, Lake Taupo, is located in the central North Island. The tourist town of Rotorua is one-and-a- half hours' driving time to the north and Tongariro National Park lies to the south of the lake. Lake Taupo has been regaled as the hub of tourism in the North Island, attracting 1.4 million visitor nights per annum, 24% of which are international visitors (Department of Conservation 2002a: 34). As a visitor destination, the lake is world renowned for trout fishing and is popular for swimming, boating, water-skiing, canoeing or kayaking and wind surfing, as well as non-water dependent recreation such as multi-sport, skiing, tramping, climbing and hunting. The township of Taupo has enjoyed strong economic growth as a domestic and international visitor destination because of the lake's resources.

For the Ngati Tuwharetoa *iwi*, the main tribe in the area, the cultural values for the central North Island are indivisible from the landscape. The sacred mountains of Tongariro National Park are one of the few World Heritage areas to have achieved the status based on natural and cultural criteria. Maori-operated tourism activities in the area include guided walks of Whirinaki and Pureora Forest Park, the newly created Wairakei Terraces and geothermal area, and several cultural performance venues. Several operators seek to provide visitors with cultural experiences on Lake Taupo. 'Te Moana Charters' is a Maori-owned and operated guided trout-fishing operation on the lake. Non-Maori-owned companies that incorporate Maori culture within their tours include 'Kayaking Kiwi', the Ernest Kemp Replica Steamboat and the Barbary sailing yacht which visit contemporary Maori rock carvings located on lakeside cliffs. These operations include Maori cultural information through retelling the legend of early Maori discovery by the chief, *tohunga* Ngatoroirangi, of the lake.

In 1992, the Maori Land Court returned legal ownership of the waters, lake bed and river tributaries of Lake Taupo to Ngati Tuwharetoa so that the tribal *mana* could be preserved and enhanced. There was a proviso that the general public would continue to have freedom of entry and access to the waters. However, the lake is managed in a partnership between Crown agencies and the *iwi* under Section 17 of the Reserves Act (1977). The Ngati Tuwharetoa Environmental Iwi Management Plan states that 'As Kaitiaki nga hapu o Ngati Tuwharetoa have an intrinsic duty to ensure that the *mauri* and therefore the physical and spiritual health of the environment is maintained, protected and enhanced' (Tuwharetoa Maori Trust Board 2001). Management issues and concerns relevant to tourism activities that were identified by Ngati Tuwharetoa included protection of *waahi tapu*; discharge of human sewage; pollution; destruction of lake habitats affecting indigenous flora and fauna;

and 'the protection of customary and traditional fishing rights and practices' (Tuwharetoa Maori Trust Board 2001).

Such issues affect other *iwi* and *hapu* (subtribe, family group) throughout the country and most recently the Te Arawa people of the Rotorua region became co-managers, with the Crown, of the Rotorua lakes. Since 1922, the Te Arawa had been paid an annuity, had ancestral fishing rights and control of Lake Rotorua's *urupa*, but limited input with management of the lakes by the Crown agencies. The annuity amounted to NZ $18,000 since 1977 but, according to Joe Malcolm, a *kaumatua* (elder) of the *iwi*, the Rotorua lakes generated '$40 million a year from wharf feed, fishing licences and so on. That money could be used to care for and manage the lakes' (Ilolahia 2001: 62). The *iwi* have only been consulted regularly about the management of the Rotorua fisheries or tourism activities on the lake since the Rotorua District Council, Environment Bay of Plenty and Te Arawa Maori Trust Board began working collaboratively to manage the lakes in the 1990s.

Visitors to Lake Rotorua have been enticed to experience the area with promotional material that alluded to the legend of the romance between a Maori maiden, Hinemoa and warrior, Tutanekai. Hinemoa dwelt on the shores of Lake Rotorua and was forbidden by her family to meet with Tutanekai. She swam the cold waters of the lake at night to avoid being detected and was united with her lover. Over the past century this popular legend has been related to many tourists who can travel to Hinemoa's bathing site on Mokoia Island. Operators invite tourists to 'trace the legend of Hinemoa and Tutanekai (a forbidden love story) and then wade through the hot pools that Hinemoa rested in after her long swim from Owhat to Mokoia Island' (Rotorua Lakes Cruises Ltd, 2004). Promotional material continues to refer to the legend, in one instance offering tourists the opportunity to 'learn the legend' and by attending a *hangi* (traditional feast cooked in an earth oven) and concert party they too may 'fall under the spell of Tutanekai's flute' (Novotel Royal Lakeside Rotorua 2003).

Fiordland National Park and Lake Te Anau

The township of Te Anau is located on the shores of Lake Te Anau at the eastern boundary of Fiordland National Park. The lake is the second largest in New Zealand, Lake Taupo being the largest. Fiordland National Park was established in 1952 and has the distinction of being part of the South West New Zealand (Te Wahipounamu) World Heritage Area. The 2001 census identified a population of 1857 people and 230 businesses in the township (Statistics NZ 2003). From Te Anau, trampers can access a number of Great Walks within the national park including the Milford, Routeburn and Kepler Tracks. The township is also a major destination for visitors en-route to Milford Sound, visitor statistics estimating approximately 133,507 visits to the Te Anau Visitor Centre in 2002 (DOC 2002b: 66).

Guidebooks and Internet websites inform visitors of the cultural values of the lakes in the area – for example:

> Tapara is the original Maori name for nearby Lake Gunn and refers to a Ngai Tahu ancestor Tapara. The lake was a stopover for parties heading to Anita Bay in search of greenstone. (Lonely Planet 2002: 728)

An Internet site relates the legend of Lake Te Anau:

> Legend has it that the lake was called 'the lake of infidelity' after the wife of the head warrior betrayed her husband by showing a neighbouring tribal warrior where the sacred well of life was and when the neighbouring tribesman drank from the well it over flowed flooding and drowning the tribe and forming what is now Lake Te Anau. (Independent Traveller's Adventure Guide 2003)

Within the Te Anau Visitor Centre, several displays refer to Ngai Tahu values. Ta Te Raki Whanoa, an ancestral figure, is the focus of one wall panel and a diorama display presents early Maori lifestyle within a cave setting. Panels within the diorama relate the importance of Fiordland to Ngai Tahu as the local *tangata whenua* and the significance of the mountain, Aoraki/Mount Cook. Other panels in the cave diorama discuss traditional clothing, food sources and domestic matters such as shelters, fire lighting and sleeping areas.

Since the late 1990s, the Department of Conservation Te Anau summer holiday programme has incorporated a Ngai Tahu perspective on the area. Between five and seven of the Te Anau summer holiday programme activities (walks and talks) incorporate 'conservation from a Maori perspective', including 'Moods of Manapouri' which is a boat cruise on nearby Lake Manapouri, guided by a Ngai Tahu interpreter. On this excursion visitors can 'Hear how the early Maori journeyed through this landscape and view some of their old resting and food gathering spots around the lakeshore' (DOC 2000).

Real Journeys, formerly Fiordland Travel, offers boat tours on Lake Te Anau. While it is the natural heritage of the area (fiords, lakes and mountains), rather than cultural features that are marketed to visitors to the region, the company incorporates local Maori (Ngai Tahu) perspectives within its tours as a result of staff participation in cultural workshops provided by Ngai Tahu. The tour to Te Anau caves is a half-day excursion commencing with a boat trip across Lake Te Anau and the boat skipper's commentary includes information about the Manapouri hydro-electricity scheme, captive rearing of *takahe* (an endangered bird) by the Department of Conservation (DOC) and early Maori or European history of the area. The skipper usually mentions the Waitaha and Ngai Tahu people on seasonal food-gathering expeditions. The commentary may incorporate the use of *mokihi* (rafts) by the Maori, which were used to cross the lake on the trip to Milford Sound in search of *takiwai o pounamu*, a type of greenstone or precious jade. These *mokihi* were also used on Lake Manapouri. At Te Anau caves there are display panels in the visitor centre relating cultural aspects of the area prior to the guided cave trip. One of the exterior panels situated by the lake and entitled 'Swirling Waters' mentions the Maori name of the Te Anau Caves. The name was originally Te Ana-au and meant 'the cave with a current of water swirling in and out', possibly forming a whirlpool. Another panel 'Mirrored in the Pool' has distinctive *kowhaiwhai* (traditional decorative artwork) pattern borders and tells a traditional legend of the area. Promotional information, presented in the 'Te Anau Glow-worm Caves' brochure produced by Real Journeys, includes a page devoted to Maori tradition entitled 'Lost in Legend', raising visitor awareness of the cultural values for the area.

Fiordland National Park tourism operations, *iwi* and national park management mindfully work towards providing visitors with opportunities to learn about the cultural values for the area. *Mahinga kai*, pre-European Maori history and stories linking the ancestors of early Fiordland Maori to the surrounding lake and alpine environments are dominant themes related to visitors through commercial and non-commercial activities.

Lake Pukaki

Lake Pukaki is located in the Mackenzie Country of South Island. The lake is world renowned for its turquoise colour and the spectacular view of the Southern Alps. Lake Pukaki and its surroundings were considerably altered in 1979 when, on completion of a major hydro-electricity scheme in the Mackenzie Country, the lake volume increased considerably upon back-filling of a dam. In terms of recreational values, the lake's glacial characteristics reduced water clarity (but this contributes to the renowned turquoise colouring) and the icy temperatures have limited its use for activities such as boating, swimming or kayaking. Recreational activity has been concentrated during the summer season at other lakes in the Mackenzie Country – for example, Lake Tekapo (water-skiing and boating), Ohau (boating and fishing), Alexandrina (fishing, boating) and Ruataniwha (boating, water-skiing, jet-skiing, rowing, swimming), as have the hydro-electricity lakes of the Waitaki Valley, including Benmore.

Lake Pukaki is one of many lakes and rivers of significant value to *tangata whenua* in the region. The land around Lake Pukaki has yielded remnants of archaeological evidence of seasonal visits by pre-European Maori. Numerous rock-art sites and *waahi tapu* areas were submerged with the rising lake level on completion of the hydro-electricity project, thus leaving few tangible remains of early Maori culture in the area for the contemporary visitor. The Ngai Tahu and Waitaha peoples have a rich oral history that records visits by those who sought *mahika kai* (Tau *et al.* 1990). The original lake area, prior to hydro-electricity development, was the site for camping settlements and had important habitat where *mahika kai*, particularly waterfowl and freshwater eels were gathered. The lake is filled by melt-waters originating from a sacred source for *tangata whenua* – Aoraki/Mount Cook, the *tupuna* (ancestor) of Ngai Tahu and the Waitaha peoples (Tau *et al.* 1990). For both the Waitaha and Ngai Tahu people the waters of Aoraki/Mount Cook were sacred and are still used for ceremonial purposes (Tau *et al.* 1990). The melt-waters from Aoraki/Mount Cook enter the lake via the Tasman and Hooker glaciers and their tributaries, the Hooker and Tasman rivers. The cultural values for the lake are typical of a widespread respect the Maori have for the *mauri*, or life-force of water, within lakes and rivers.

With the Ngai Tahu Treaty Settlement Act (1998) a *nohoanga* was mapped for members of Ngai Tahu on the eastern shores of Lake Pukaki so that members of the *iwi* could make annual trips to the area and camp. While affording superb views of Aoraki/Mount Cook, the site is also exposed to strong prevailing north-west and southerly winds, thus making it at times a challenging and inhospitable area in which to camp. This is but one of several *nohoanga* located beside lakes and rivers in the Mackenzie Country. The *nohoanga* have been controversial at times as there is a policy of no public camping, much to the ire of locals and visitors to the area. In 1998, members of the Otago Conservation Board objected to the designation of *nohoanga* in the Otago region, viewing such designation as potentially causing conflict between traditional *pakeha* recreational users and Maori customary users (*Otago Daily Times* 1998). The *nohoanga* is located within a larger camping area still available for general public use. The Ngai Tahu Maori Community Law Centre responded that personal opinions of board members had failed to recognise the need for *nohoanga* so 'children may be taken to visit and experience a place that was once used by their Ngai Tahu forebears. *Nohoanga* are thus more than places to fish: they are places of cultural significance' (*Otago Daily Times* 1998: 7). Members of the *iwi* can reserve camping at the site through the Ngai Tahu centrally operated '0800-nohoanga' and all *iwi* members must fill in an authorisation form to camp in the area. This necessary management technique, which some opponents may see as bureaucratic, is an example of Western administration systems being adopted by an *iwi* authority catering for modern Maori lifestyles. Other regulations include limiting group size (usually a maximum of 30 people per night) and time of year – the area can be used from August to April, traditional food-gathering months.

There is a range of published and Internet material available to visitors concerning Lake Pukaki and the range of values related to *tangata whenua* – for example: 'Lake Pukaki was named by Rakaihautu, probably for the outlet's "swollen neck". It is the sacred resting place of the bow piece of the waka (canoe) "Mahunui" and "Mahuru"' (Tourism Waitaki 2003).

The information usually focuses on place names and legends rather than on history and contemporary issues. DOC and Meridian Energy have a formal relationship with Ngai Tahu, the latter in particular recognising the *iwi* values for the water from the rivers and lakes that is used for electricity generation. Both organisations supported the inclusion of the two pieces of interpretive material regarding the cultural values of Ngai Tahu in the Lake Pukaki information centre located a kilometre south of the *nohoanga* site. No information is yet available about the many rock-art sites submerged under Lake Pukaki, nor is any mention of the *nohoanga* made at the visitor centre. Lake Pukaki Information Centre was opened in 2000. This centre is staffed during the summer months, primarily for the purpose of providing information on activities and services at other locations in the Mackenzie Country. Approximately 90,000–100,000 visitors are recorded at the site each year attracted by the spectacular views of Aoraki/Mount Cook at the end of the lake. Next to the information centre is a kiosk that houses six interpretive panels comprising text and images, two of which present the Maori cultural values for Lake Pukaki and Aoraki/Mount Cook. One panel relates the legend of Aoraki/Mount Cook. The second panel presents the 'He Poki mo Aoraki' lament which explains the importance of Lake Pukaki and the other rivers and lakes of the Mackenzie Country in relation to the melt-waters of the snow and ice from Aoraki. The information on Aoraki was produced by DOC staff in collaboration with Ngai Tahu Rananga o Arowhenua members, and the panels have English and Maori translations. 'Close liaison with Arowhenua *rananga* and Sir Tipene O'Regan was required to ensure the translation of the legend was correct.' (Corbett 1996: 1).

No other interpretive material exists for visitors to the area regarding who the Ngai Tahu are and whether the lake has any other value for the *iwi*, despite the *nohoanga* site being so nearby. This lack of information could be regarded as non-confrontational – its absence will not provoke people who may oppose treaty settlements where land areas such as *nohoanga* are set aside for the sole use of *iwi* members. However, the two existing panels lack context as to the contemporary relationship of Maori to the area. Merely relating the legend of Aoraki and the importance of Pukaki raise more questions for the visitors, some of who may investigate the relationship further but a shortage of contextual information could also be frustrating for those visitors who lack understanding of the area.

Conclusion

In New Zealand there are increasing examples of Maori *iwi* being actively involved in the management of lakes and other natural areas visited for tourism and recreation purposes. Such involvement can assist with the protection of the *mana*, or integrity of the cultural values, located within these areas. For Maori, enhancing the environmental quality of New Zealand lakes is necessary to ensure the conservation of cultural values, particularly with regard to *mahinga kai* and *waahi tapu*. However, there is a lack of rigorous research that monitors tourism or recreational activities and associated impacts, particularly in relation to resources that are the focus of cooperative management of lakes by the Crown and local *iwi*. Such research is required to inform policy that affects the sustainable management of New Zealand lakes.

Whether there is sufficient visitor demand to enable commercial tourism organisations to operate as specialists in interpreting cultural values for lakes is debatable. The interpretation of cultural values for lakes may add value to the visitor experience; however, visitors are predominantly interested in recreational activities such as walking, fishing, boating and generally sightseeing at New Zealand lakes. The relationship of Maori to the surrounding environment is incidental to the primary motivations of many visitors travelling through New Zealand's landscape. Research and reports on tourism in New Zealand suggest there is an increasing demand for Maori cultural experiences (NZTB 1996; Ateljevic & Doorne 2002; McIntosh *et al.* 2000; Van Aalst & Daly 2002; Tourism New Zealand 2003a, 2003b; Carr 2004). Domestic visitors have been identified as major segments for some Maori tourism operations (McIntosh *et al.* 2000; Warren & Taylor 2001). Other studies indicate that international visitors are more interested in Maori culture than domestic visitors, the latter group possibly being too familiar with the culture to be interested in commercial cultural experiences (Ryan 2002; Ryan & Pike 2003; Colmar Brunton 2004). In a 2000 survey of 472 visitors to three natural areas in the southern South Island, 183 respondents (39.7%) were 'very interested' or 'extremely interested' in opportunities to learn about or experience Maori culture (Carr 2004). The respondents regularly associated Maori with Rotorua and parts of Northland, including the Bay of Islands and Cape Reinga, and placed a high level of importance on the protection of Maori cultural and spiritual values at significant sites (Carr 2004). Mythology and legends were the cultural aspects visitors were most interested in (n = 134), followed by environmental knowledge or the use of natural resources (n = 115). The Colmar Brunton report, *Demand for Maori Culture*, indicated that 53% of travellers interviewed (n = 666) were interested in visiting sites of cultural significance to Maori and 38% had actually visited such sites (Colmar Brunton 2004).

It can be concluded that New Zealand lake destinations provide more than just scenic settings or recreational opportunities for visitors. The treaty settlements of the past decade have apparently facilitated opportunities for Maori themselves to participate actively in lakes and water-based recreation. The designation of campsites or *nohoanga* has encouraged the revitalisation of activities such as gathering *mahinga kai*. Existing visitor centres and interpretation around lake areas can incorporate information to educate visitors about the unique cultural values that Maori have for such environments (O'Regan 1990; Keelan 1996; Carr 2004). The provision of *iwi*-owned and operated businesses or the development of interpretation about the diverse cultural values for lakes may enable visitors to have cultural experiences, while strengthening the identity and economic well-being of Maori communities.

References

Aotearoa Maori Tourism Federation (AMTF) (1994) *Position Paper on the Protection of Maori Cultural and Intellectual Property within the Tourism Industry*. Rotorua: AMTF.

Ateljevic, I. and Doorne, S. (2002) Representing New Zealand: Tourism imagery and ideology. *Annals of Tourism Research* 29 (3), 648–67.

Avery, P. (1999) Between myth and history: Mythology and legends as tourist attractions. In E. Arola and T. Mikkonen (eds) *Tourism Industry and Education Symposium Proceedings, Reports and Proceedings* (pp. 140–58). Jyvaskyla Polytechnic No. 5, Jyvaskyla Polytechnic.

Brown, T.J. (1999) Antecedents of culturally significant tourist behaviour. *Annals of Tourism Research* 26 (3), 676–700.

Carr, A.M. (2004) Mountain places, cultural spaces – interpretation and sustainable visitor management of culturally significant landscapes. *Journal of Sustainable Tourism* 12 (5), 432–59.

Colmar Brunton (2004) *Demand for Maori Tourism*. Wellington: Ministry of Tourism.

Corbett, R. (1996) Lake Pukaki Information Kiosk. *Canterbury Conservancy News*. May (p. 2). Christchurch: Department of Conservation.

Department of Conservation (2000) 2000–2001 Summer Holiday programme. Unpublished. Te Anau: Department of Conservation.

Department of Conservation (2001) *Visitor Strategy*. Wellington: Department of Conservation.

Department of Conservation (2002a) *Tongariro: The Annual Journal of the Tongariro/Taupo Conservancy* (Summer). Turangi: Department of Conservation.

Department of Conservation (2002b) *Draft Fiordland National Park Management Plan*. Invercargill: Department of Conservation.

Devlin, P. and Booth, K. (1998) Outdoor recreation and the environment: Towards an understanding of the recreational use of the outdoors in New Zealand. In H.C. Perkins and G. Cushman (eds) *Time Out? Leisure, Recreation and Tourism in New Zealand and Australia* (pp. 109–26). Auckland: Longman.

Hall, C.M. and McArthur, S. (1996) Managing community values: Identity – place relations: An introduction. In C.M. Hall and S. McArthur (eds) *Heritage Management in New Zealand and Australia* (pp. 180–4). Melbourne: Oxford University Press.

Hinch, T. (1998) Tourists and indigenous hosts: Diverging views on their relationship with nature. *Current Issues in Tourism* 1 (1), 120–4.

Ilolahia, N. (2001) They're darling those places. *Mana* October/November 2001, 60–3.

Independent Traveller's Adventure Guide (2003) Available at www.itag.co.nz/teanau.asp. Accessed April 2003.

Ingram, T. (1997) Tapoi Tangata Whenua: Tapoi Maori ki Aotearoa – Indigenous Tourism: Regional Maori Tourism in Aotearoa. In J. Higham and G. Kearsley (eds) *Proceedings Trails in the Third Millenium Cromwell, Otago, New Zealand.* Dunedin: Centre for Tourism, University of Otago.

Kearsley, G.W., McIntosh, A.J. and Carr, A.M. (1999) Maori myths, beliefs and values: Products and constraints for New Zealand Tourism. In E. Arola and T. Mikkonen (eds) *Tourism Industry and Education Symposium Proceedings, Reports and Proceedings.* Jyvaskyla Polytechnic No. 5 (pp. 292–302). Jyvaskyla Polytechnic.

Keelan, N. (1996) Maori Heritage: Visitor management and interpretation. In C.M. Hall and S. McArthur (eds) *Heritage Management in New Zealand and Australia* (pp. 195–201). Melbourne: Oxford University Press.

Lomax, H.M. (1988) Maori use and non-use of national parks. Master of Arts thesis, University of Canterbury.

Lonely Planet (2002) *New Zealand.* Compiled by P. Harding, C. Bain and N. Bedford. Melbourne: Lonely Planet Publications.

McIntosh, A.J., Smith, A. and Ingram, T. (2000) *Tourist Experiences of Maori Culture in Aotearoa, New Zealand.* Dunedin: Research Paper Number Eight, Centre for Tourism, University of Otago.

MacLennan, P. (2000) *Visitor Information as a Management Tool.* Science and Research Internal Report 180. Wellington: Department of Conservation.

Matunga, H. (1995a) *Maori Recreation and the Conservation Estate.* Centre for Maori Studies and Research: Lincoln University.

Matunga, H. (1995b) Maori participation in outdoor recreation: An exploration of the research and issues. In P.J. Devlin, R.A. Corbett and C.J. Peebles (eds) *Outdoor Recreation in New Zealand Volume 1: A Review and Synthesis of the Research Literature.* Wellington: Department of Conservation and Lincoln University.

New Zealand Tourism Board (NZTB) (1996) International Visitor Survey. Wellington: New Zealand Tourism Board.

Novotel Royal Lakeside Rotorua (2003) Novotel Royal brochure.

O'Regan, S. (1990) Maori control of the Maori heritage. In P. Gathercole and D. Lowenthal (eds) *The Politics of the Past* (pp. 95–106). Cambridge: Cambridge University Press.

Otago Daily Times (1998) OCB response 'inflammatory', 28 May, 7.

Pfister, R. (2000) Mountain culture as a tourism resource: Aboriginal views on the privileges of storytelling. In P.M. Godde, M.F. Price and F.M. Zimmerman (eds) *Tourism and Development in Mountain Regions* (pp. 115–36). Wallingford: CABI Publishing.

Rotorua Lakes Cruises Ltd (2004). Available at www.ubd.co.nz/company-profiles/2067443097.

Ryan, C. (2002) Tourism and cultural proximity: Examples from New Zealand. *Annals of Tourism Research* 29 (4), 952–71.

Ryan, C. and Pike, S. (2003) Māori-based tourism in Rotorua: Perceptions of place by domestic visitors. *Journal of Sustainable Tourism* 11 (4), 307–21.

Shackley, M. (2001) *Managing Sacred Sites.* London: Continuum.

Sinclair, D. (1992a) Land: Maori view and European response. In M. King (ed.) *Te Ao Hurihuri: Aspects of Maoritanga* (pp. 65–84). Auckland: Reed Books.

Sinclair, D. (1992b) Land since the Treaty. In M. King (ed.) *Te Ao Hurihuri: Aspects of Maoritanga* (pp. 85–105). Auckland: Reed Books.

Southland Times (2000) PM not keen on racist camps, 23 August, 3.

Staiff, R., Bushell, R. and Kennedy, P. (2002) Interpretation in national parks: Some critical questions. *Journal of Sustainable Tourism* 10 (2), 97–113.

Statistics New Zealand (Te Tari Tatau) (2003) *Community profiles: Te Anau*. Available at www.statistics.govt.nz.

Tau, T.M., Goodall, A., Palmer, D. and Tau, R. (1990) *Te Whakatau Kaupapa: Resource Management Strategy for the Canterbury Region*. Christchurch: Aoraki Press.

Tourism New Zealand (2003a) *Tourism New Zealand Three Year Strategic Plan 2003–2006*. Wellington: Tourism New Zealand.

Tourism New Zealand (2003b) *New Zealand's Ideal Visitor: The Interactive Traveller*. Wellington: Tourism New Zealand.

Tourism New Zealand (2004) Media release 'Five Great Years'. Wellington: Tourism New Zealand.

Tourism Waitaki (2003) *Beautiful Waitaki: Ocean to Alps* – Available at http:// beautiful-waitaki.co.nz/background.htm. Accessed May 2003.

Tuwharetoa Maori Trust Board (2001) *Taupo-nui-a-Tia 2020*, Tuwharetoa Maori Trust Board: Turangi.

Upitis, A. (1989) Interpreting cross-cultural sites. In D. Uzzell (ed.) *Heritage Interpretation: Volume 1, Natural and Built Environment* (pp. 142–52). New York: Belhaven Press.

Van Aalst, I. and Daly, C. (2002) International visitor satisfaction with their New Zealand experience: The cultural tourism product market – a summary of studies 1990–2001. Unpublished report. Wellington: Tourism New Zealand.

Walker, R. (1992) The relevance of Maori myth and tradition. In M. King (ed.) *Te Ao Hurihuri: Aspects of Maoritanga* (pp. 171–84). Auckland: Reed Books.

Walker, R. (1996) Contestation of power and knowledge in the politics of culture. *He Pukenga Korero Ngahura* 1 (2), 1–7.

Warren, J.A.N. and Taylor, C.N. (2001) *Developing Heritage Tourism in New Zealand*. Wellington: Centre for Research, Evaluation and Social Assessment (CRESA).

Part 3: Tourist Activities and Perceptions

Chapter 6
Lakes as an Opportunity for Tourism Marketing: In Search of the Spirit of the Lake

ANJA TUOHINO

Introduction

In the development of a tourist destination, an important role has long been played by marketing, which probes the wishes and likings of tourists. In order to understand them, marketing people have to know what they are. All too often marketing strategies are created without proper knowledge of what the tourist really wants. Supply and demand don't meet if they are dealt with as separate entities independent of each other. Marketing organisations create their own strategies and the tourism providers their own products without co-operating with one another. In particular, marketing based on special products or special interests (e.g. island tourism, alpine tourism, adventure tours) has developed on the basis of individualised destination marketing (Walsh *et al.* 2001). As for lake tourism, destination marketing has only barely started.

The images and travel experiences portrayed in publicity materials should reflect the tourists' motivations for travel. For the travel agency, a trip is a saleable product consisting of various services, while for a tourist planning one, it is an immaterial combination of future experiences (Sepänmaa 1998). However, the question arises how commercialisation and marketing are able to bring out the 'spirit' or sense of place of a region. Can a mere mental image of a place limited by the sense of vision and the angle of view of the photographer arouse the motivation to travel to a destination that can be experienced more fully through all senses? Can the *genius loci*, the various associations connected with a place, be captured in the pictorial communication of tourism marketing? These are the questions that the author is trying to answer in this chapter by aiming at bringing new ideas and angles of vision to bear on the analysis of tourism marketing and publicity materials, as well as on the planning of the content of the messages conveyed through marketing.

Lakes are generally considered the main tourist attraction of Finland (see e.g. Vuoristo & Vesterinen 2001; see also Vepsäläinen & Pitkänen, Chapter 4, this volume), but in the marketing of tourism, promotional pictures have long repeated the same established angles and motifs that emphasise the wilderness point of view, forgetting the local touch. This study aims at showing that introducing the concept of 'sense of place' into the mental landscapes connected with the lake allows the development of lakes as a tourism resource and the touristic development of a region. Raising the 'sense of the lake' to the ranks of traditional tourism images increases the value of lake tourism and lake landscapes. This chapter provides a possible 'foreign' interpretation of the value of lake tourism – in this case, the German and Italian interpretation of the 'spirit of the lake' sensed by the respondents and finding the mental images through which one can more easily grasp the conceptual and symbolic character of the lake district.

The Lake as a Place and as a Landscape

The built-up and natural physical environment of a destination, its local culture and community spirit can be considered as the core resources in tourism. From the resource point of view, the attractions have a vital importance. The attractiveness of a destination as a marketing resource can be considered from different angles, such as those of nature and landscape, climate, culture, history, the possibility of engaging in various hobbies and accessibility. However, the resource itself is not a product. Rather, existing resources are the necessary precondition for the creation of a travel experience that can be turned into a saleable tourism product (Järviluoma 1994; Middleton 1997; Middleton with Clarke 2001). The marketing of the Finnish lake regions mainly rests on nature and landscapes, since our lakes are sold with an emphasis on virgin forest. In the Western world, the wilderness is historically part of 'otherness' to us. It has been *terra incognita*, a strange region outside the sphere of western civilisation. In the last few years, however, it has quickly come into public awareness as a tourism resource and attracted a growing amount of research (e.g. see Hall & Page 1999).

This study approaches the lake landscape of tourism marketing from the viewpoint of humanistic geography. Where the more traditional area holistic approach interprets the landscape as a measurable part of the physical environment, humanistic geography presents a multi-layered view that goes beyond what we see, interpreting the landscape as a more comprehensive visual and experiential phenomenon created through our mind in addition to sensory observations. The landscape, then, is the totality of the knowledge and feelings particular to each observer (Paasi 1986; Saarinen 1995; Raivo 1997).

The word 'landscape' is polysemous. Geographical landscape research has studied both traditional entities based on area classification and on the physical and cultural essence of the landscape, as well as our experience of the landscape. Initially, what we see, hear and smell are the foundations of our observed landscape. An observation, however, cannot be a mere objective physical experience. It is created only through our interpretation and is therefore inseparable from the latter (Kanninen 1993; Karjalainen 1995). A lake landscape becomes concrete in action through bodily experience or sensory observations in, for instance, boating or swimming. As an experience space, the lake landscape is defined as forest, shores and trees, and it is spoken about through experiences, memories, feelings and activities.

The representative landscape is one that contains signs of a given communicative system referring to space. The landscape can be seen as a manuscript containing the natural history of the region. Similarly, a cultural landscape can be seen as the totality of its physical features and the traces left by human activity, including language, food and clothing, as well as politics, religion, architecture, settlement and even population. *How* we see the elements of the landscape, not *what* elements the landscape contains, is important to the interpretation of a cultural landscape (Hudman & Jackson 1990; Park 1994; Raivo 1995b).

'Landscape' and 'place' are both concepts through which humans act in relation to their environment. A landscape can be looked at by an experienced participant or by an outsider observer, while a place takes on meaning through the experience and interpretation of an individual (Meinig 1979). In traditional tourism marketing, the lake environment has been simplified as wilderness – a wild and free natural landscape. This point of view, however, ignores the landscape as object of experience and subjective interpretation. A neutral lake environment becomes a meaningful place after the tourist links to it mental images and feelings formed through experience. A given place becomes part of the person when she feels she belongs to it and makes it her home.

The concept of place is often connected with the adjective 'safe', but negative feelings may be part of a place. Tuan (1974) defines 'topophilia' as a strong sense of belonging to a place, feeling the place as one's own. 'Topofobia' in turn refers to negative feelings such as aversion or fear. According to Tuan (1979), a strong feeling such as fear connected with a place is in the mind as much as in the environment. A sense of fear in a space is the subjective equivalent of an objective and unknown environment. Similarly, placelessness refers to an environment without important places, as well as to an attitude that does not recognise the meanings of places (Relph 1976; Tuan 1977, 1979; Karjalainen 1986; Haarni *et al.* 1997).

'Place is security, space is freedom.' We are tied to one and long for the other. Travel is movement from one place to the next in space. Travel then responds to an individual's longing for freedom, separating him from his everyday environment, while at the same time it is connected with hopes and expectations that the destination should become a meaningful place. People look for places that differ from their usual surroundings physically as well as socially. On the other hand, they look for the secure and the well-known, for a 'home away from home'. A trip is an experience of movement in which time and space are commercially utilised, even if in our thoughts and feelings we are transferred into something deeper and more mythical (Tuan 1977; Crompton 1979; Hemmi & Vuoristo 1993; Tuohino & Uusi-Illikainen 1997).

According to Raivo (1995a), a landscape is a subjective interpretation – a landscape of mind that varies from one individual and one group of people to the next. Landscapes of mind are born not only from personal experiences and the sensory observation of the environment, but also from indirect spatial representations such as cultural products. The representations of landscapes and places in commercial tourist brochures also tell about their essence, their symbolic and cultural importance. A mental trip by means of representations is, in fact, tourism in space, time and place (see Tuohino & Uusi-Illikainen 1997).

The Sense of Place and Experiencing It

One of the important concepts of humanistic geography is 'sense of place'. This concept was introduced to geography by Tuan in the 1970s. The sense of place has been interpreted as a social concept and as an individual value or phenomenon. The concept of 'sense of place' has gained ground in tourism research in recent years. It refers to a positive sense of oneness of a person with a place he interacts with. To quote Tuan (1974), 'people demonstrate their sense of place when they apply their moral and aesthetic discernment to sites and locations'. A place is a piece of reality exemplified and interpreted through meanings given to it, while 'sense of place' can be equated with the identity of a place (Tuan 1974, 1979; Relph 1976; Eyles 1985; Shamai 1991; Hall & Page 1999; Meyer 2001; Moisey & McCool 2001; Walsh et al. 2001).

'Sense of place' meaning 'the associations connected with a place' is important to the development of tourism for more than one reason. The sense of place is important both to the tourist and the developer of tourism, since it represents what is unique and worth preserving about a place. First, if we understand the sense of the place, the *genius loci*, we can adopt the place and locality as the core marketing points and pay less attention to general features. Second, this approach recognises the destination as a node in the network of neighbouring regions, thus giving rise to a more holistic sensation of place. Third, understanding this concept brings the marketing effort to the local level and commits the local population to experiencing the destination through commercialisation and local development. Although the physical resources are the starting point of tourism, the importance of social and cultural resources should not be neglected, since it is the latter that represents the identity and meanings of the place. Through its local knowledge and expertise, the population of the place is a potential resource that can create the spirit of the place. Forgetting the local touch in marketing strategy may lead to a distortion of the images marketed and reality does not come up to the expectations of the tourist (Meyer 2001; Walsh *et al.* 2001).

The present research aims at finding the sense and spirit of the place from another viewpoint. Answers are sought to questions such as 'can the tourist find or experience the spirit of the place through photos used in marketing without direct personal interaction or attachment to a place'. 'Can the familiar be found in a unfamiliar place?' 'Can an unfamiliar place be full of meanings?'

The Spirit of the Lake

If we approach the lake through the concept of 'spirit of the lake', we first have to consider the lake as a place or (at least) as a meaningful landscape. A lake is more than forms of land and water or the play of sunlight on the waves. Entering the landscape and experiencing it raises into consciousness many experienced feelings and old memories.

> Those sitting on the shore, the swimmers, rowers, anglers, skiers on the ice, are all experiencers and doers alone and together, all in their own ways seeking unity with the lake and the landscape of the lake. (Lehtinen 2001; Koskela 2002)

Water has always been important to people. Waterscapes also are important as tourism attractions (Tuohino & Pitkänen 2004a). Water is a different element from land and takes a central role in many myths pondering the shape of the world. Water is essential and, at the same time, mysteriously frightening. Beside its mythical history, the Finnish lake landscape also has its roots in natural history and culture. It is a unique monument of the Ice Age, whose labyrinth-like structure and rocky shores are signs of the glacier that melted away thousands of years ago. Later, humans came to the same shores with tools and animals. The early communities of hunter-gatherers and farmers gave way to more efficient methods of utilising and conserving the lake environment (Lehtinen 2001). The lake landscape has also many symbolic meanings in Finnish culture. During the rise of the nationalist ideology at the turn of the 19th century, the lakes were adored as a national landscape.

Correspondingly, ever since the beginning of urbanisation in the 1950s, the lake landscape has been seen as a symbol of the 'golden youth' and countryside nostalgia (see Vepsäläinen and Pitkänen, Chapter 4, this volume). It has become almost 'a must' to invest in one's own summer cottage by a lake (Pitkänen & Vepsäläinen 2003). Today, Finnish lakeshores are dotted with summer cottages and holidaymakers and local people alike enjoy the lake landscape in the form of various kinds of activities (Tuohino & Pitkänen 2004b).

Hence, the lake environment is structured as the lake landscape comprehended through our thinking that utilises its historical forms and deeper individual and cultural meanings. Sensing the landscape imbues it with a feeling of familiarity that in turn reinforces the experience.

> You can meet your lake over and over again, different but still similar, where it should be, in the same place. (Lehtinen 2001)

As experienced space, the lake landscape is defined as sensual perceptions and their interpretation as water, shores and forests. In the same way, the landscape is talked about not only through experiences, memories, feelings and activities, but also through the cultural interpretations, meanings and sign systems connected with them. Through human interaction with nature. lake landscape becomes a 'lived world' full of values and meanings, putting it somewhere between culture and wild nature (Tuohino & Pitkänen 2004a; see also Tarasti 1990; Suvantola 2002).

The Methods, Data and Analysis of the Study

This study uses pictorial analysis to capture the images evoked in potential tourists to Finland by marketing pictures showing a lake landscape. The research data was generated using ten marketing pictures used by the Finnish Tourist Board (FTB). The pictures used were chosen in co-operation with representatives of the FTB. An attempt was made to choose pictures showing a large variety of the tourist potential offered by the lake resource. Besides showing a typical lake landscape (often a lake and forest), a picture had to look recognisably Finnish. The pictures were part of the FTB series 'Sommermotive' and were intended to be used in marketing abroad. The water element was prominently shown in all the pictures.

The target group consisted of Germans and Italians, all of whom were interviewed individually at international tourism fairs in Germany and Italy in February and March 2002. Interviews were carried out in the respondents' native language. In Germany, the 60 persons interviewed were selected by a random sampling method from among visitors to the fair, while in Italy 50 respondents were tourism professionals working at the stands. In Italy, the interviews were made on 'pros only days' and in Germany on 'open to public days'. The choice of the groups was based merely on practical reasons (access to the fair and time available). Therefore, it is assumed that there is not a notable difference among respondent groups in evaluating the photographs. Thus, the answers of tourism professionals did not differ essentially from the answers of the other visitors.

None of the interviewees, with two exceptions, had visited Finland, so they had no personal connections with the topics shown in the pictures. As for those who have visited Finland, familiarity with the topics influences their mental images, since their knowledge of the target is based on facts, unlike that of the bulk of the respondents. The interviewees were shown the pictures one by one, so they could not form any advance idea. In the interview situation, the interviewees were asked only one question: 'What kind of impressions do you get from these pictures?' and the responses were recorded on a form. Interviews took about five minutes per person and the results show their spontaneous impressions from the pictures shown.

The emphasis was on image components studied using an unstructured form with open-ended questions. The attributes found were assigned to categories through which the tourism potential of the lake could be approached. To avoid influencing the respondents, no ready-made alternatives were given on the form. An open question was used in an attempt to find the images the respondent connected with the picture. In a later categorisation, the open responses were linked to the pictures. The answers to the mental image question were analysed using the normal classification and categorisation methods of qualitative research. The NVivo programme suitable for the analysis of qualitative data was used to examine the data.

The data was analysed by inductive analysis. Only the unequivocal messages of the respondents were taken into account. The answers were classified and assigned to groups. Before classification, the data was checked several times in order to form a general opinion. Classification proceeded in bottom-up direction – i.e. each word/expression was coded, after which the groups were combined to find more general categories. Five general categories were arrived at: functional-social, physical, aesthetic, symbolic and cultural images. Attributes connected with activities and socialising were assigned to the 'functional-social' category, those connected with landscape forms to the 'physical' category, while all attributes connected with the layout, colours or composition of the pictures were assigned to the 'aesthetic' category. Attributes referring to traditions and culture were assigned to the category 'cultural' and attributes with a symbolic content, to the category 'symbolic'.

The main principle applied in coding was to use labels that are semantically transparent. The number of codes was ignored at this stage and only words formed on the same root (e.g. love *n* and love *v*) were coded similarly. At the next stage, root-level codes were assigned to a given subcategory on the basis of semantic features (e.g. quiet *n* and relaxation *n* were assigned to the subcategory of peace and rest). At the final stage, the subcategories were coalesced under the main mental image categories: 'functional-social', 'physical', 'aesthetic', 'cultural' and 'symbolic'. The main categories follow Echtner and Ritchie's well-known four-field model commonly applied in image research (see Echtner & Ritchie 1993).

Outsider or Participant in the Landscape of Marketing Images?

The mental images evoked by the pictures in this study are both individual and social at the same time. Although the individual nature of the mental images became prominent in the analysis, there were identifiable similarities. In the analysis, each picture proved to be functional from a marketing point of view, in a given country. In the following an analysis of the spirit of the place through the viewing of three different pictures will be discussed. Table 6.1 shows how responses are divided into main categories and subcategories between Germans and Italians.

In the picture 'Cyclists', two cyclists have stopped on a summer day to enjoy the lake landscape and a sunny atmosphere. The picture clearly emphasised the intercultural differences. The Germans (Figure 6.1) strongly identified with the cyclists in the picture and the images evoked by the picture were connected with hiking and simultaneous observation of nature. Like the Germans, the Italians (Figure 6.2) in their responses emphasised various activities with the difference that nature, rather than hiking, was seen as the target of activity. Making contact with nature meant inner peace and a sense of freedom for the Italians. Both groups sought to find the inner meanings of the landscape through their respective images – the Germans through the experience created by activity and observation and the Italians through the emotions aroused by the same things. In both groups, some experienced the image of the photograph as stressful, difficult and uneasy.

To the German respondents, the environment of the picture 'Angling' was connected with peace and relaxation (Figure 6.3). The special relationship between father and son was identified as a family idyll cherished in an environment apt to match the relationship. For the Italians too, the picture of an angler primarily meant peace and relaxation (Figure 6.4). Whereas the Germans saw the picture through the value-related feelings aroused by it, the Italians connected the picture with place-related feelings. The landscape of the picture was seen as 'familiar' and memories were linked to it. The topophilic feelings associated with the place produced a soothing effect. For the German viewers the lake picture

Table 6.1 Responses divided into main categories and subcategories

	Picture 1		Picture 2		Picture 3	
	Germans	Italians	Germans	Italians	Germans	Italians
Functional-social	52	59	80	67	22	9
• Holiday	8	5	7	5	1	
• Activities	24	27	14	11	5	
• Stressfulness, restlessness	7	7			3	
• Peace, relaxation	8	10	40	49	6	3
• Social activities	3	6	19		4	3
• Way of life				2		
• Experiences	1				3	3
• Health	1	4				
Aesthetic	23	22	10	26	41	49
• Feelings	4	15	2	18	9	39
• Picture composition	4	2	3	2	4	2
• Beauty	13	5	4	6	9	2
• Idyllic	1		1		5	
• Atmosphere						3
• Unnatural	1				0	
• Juxtaposition					2	3
• Romance					12	
Physical	25	19	10	7	25	30
• Elements of nature	17	5	7	6	12	18
• Countries and places	2		2		6	
• Landscape		7				
• Characteristics of nature	6	7	1	1	3	12
• Time					4	
Symbolic					3	
• Mysticism, rituals					3	
Cultural					9	12
• Tradition, burying, Easter bonfire					9	
• Indians, aboriginal, the places of antics						12

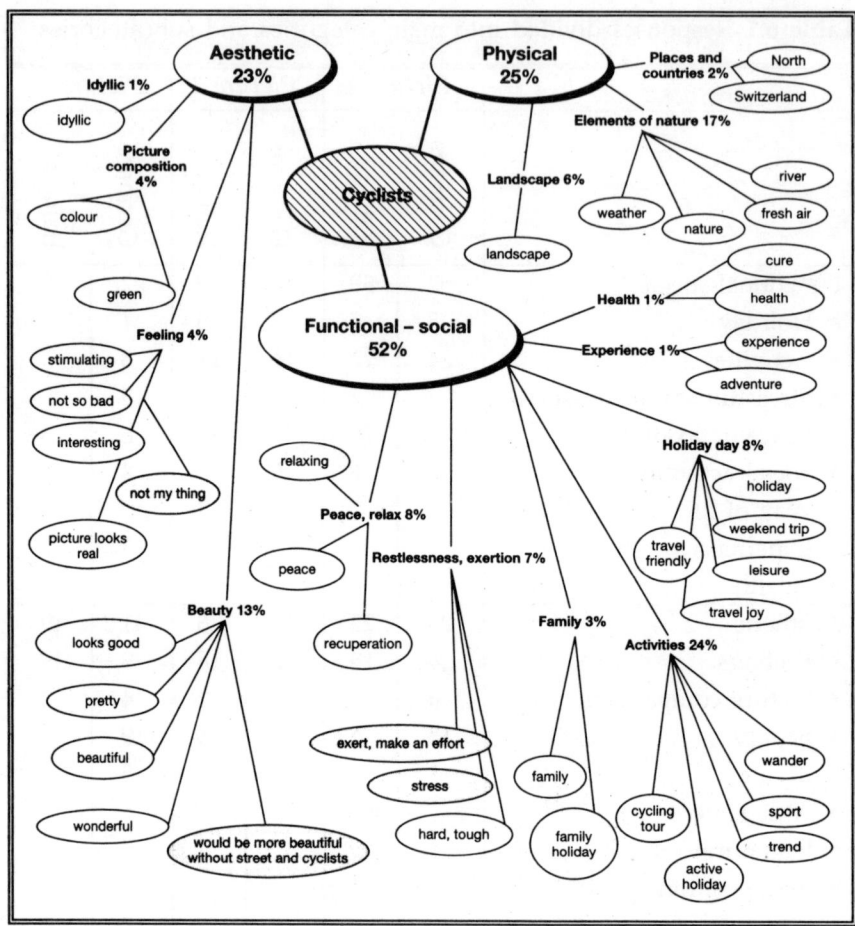

Figure 6.1 German respondents' comments on picture 'Cyclists'

with the angler mainly represented social interaction, whereas for the Italians it was primarily a geographical location endowed with emotion.

The reception given to the picture 'Midsummer Bonfire' was the most controversial among both Germans and Italians. The Germans (Figure 6.5) expressed the most positive attitudes towards the picture, associating the picture with the mystique of the opposite elements fire and water. The Italians, too, approached the picture 'from the inside' (Figure 6.6). The bonfire was more unfamiliar to the Italians than to the Germans and evoked negative feelings to a greater extent. The man-made fire in the picture was regarded as a dangerous element out of place in a natural landscape.

In Search of the Spirit of the Lake 111

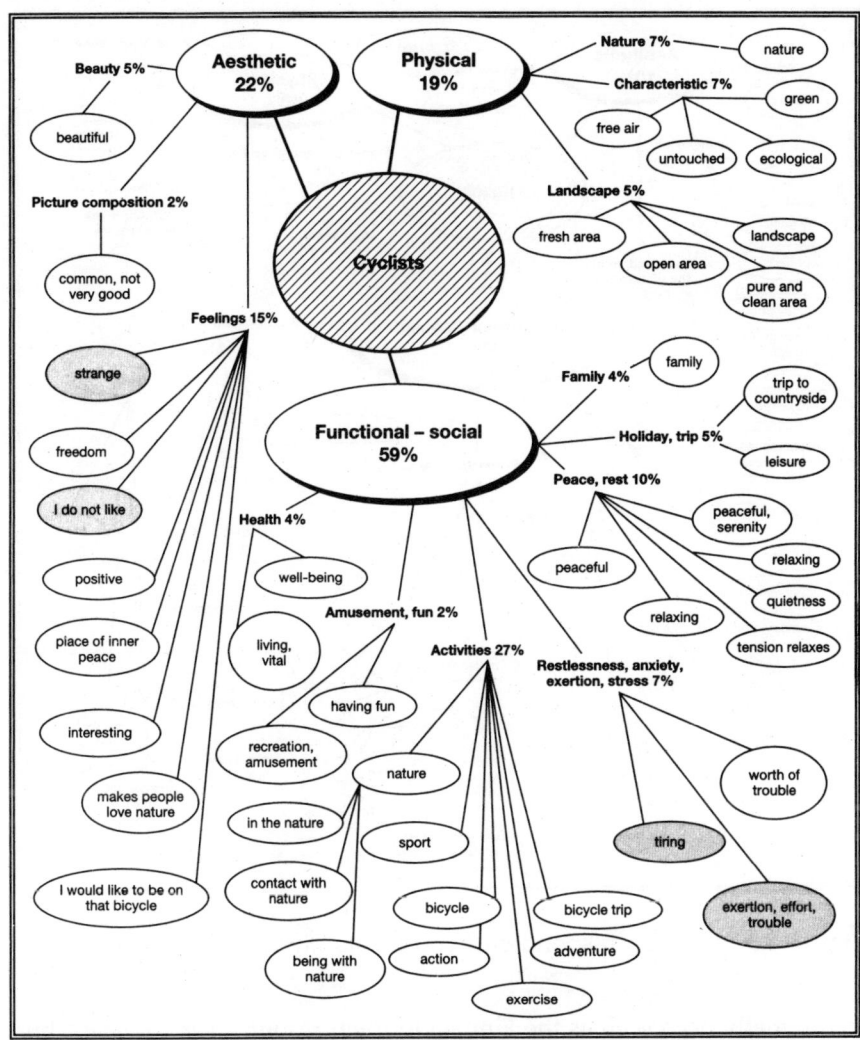

Figure 6.2 Italian respondents' comments on picture 'Cyclists'

The Meanings of Mental Images and Their Challenge to Tourism Marketing

The demands tourists are making on destinations are increasingly diverse. The trend is towards new experience and adventure, which are sought through culture as well as activities. Tourists are choosing more and more individualised destinations, which set new demands on marketing in terms of personal touch and memorability. Lakes have

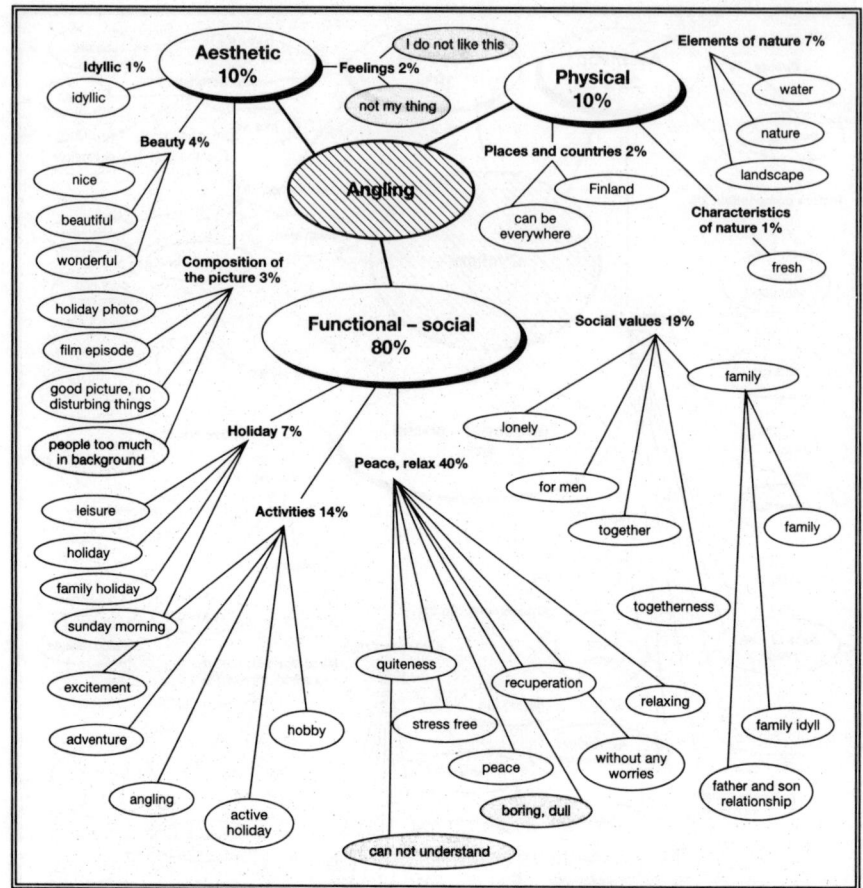

Figure 6.3 German respondents' comments on picture 'Angling'

traditionally been seen as the Finnish tourism resource *par excellence*, but it would not be too wrong to say that the marketing of the lake region has kept on repeating the ideals of Finnish Romanticism for more than a century (see Vepsäläinen and Pitkänen, Chapter 4 this volume). A mosaic of water and islands fading into the horizon is not in itself enough for the modern tourist avid for experiences.

As described, lakes must be made 'alive' by giving them a meaning that is understandable to the target audiences. Marketing a lake landscape involves finding out what kind of meanings the landscape can and even more importantly, cannot impart to the target audiences. In marketing, there is also a need to tell tourists what they can do on the lake or near the lake (see Tuohino *et al.* 2004). Certainties are culturally biased; the iconic meanings of the lake landscape are obvious only to

In Search of the Spirit of the Lake

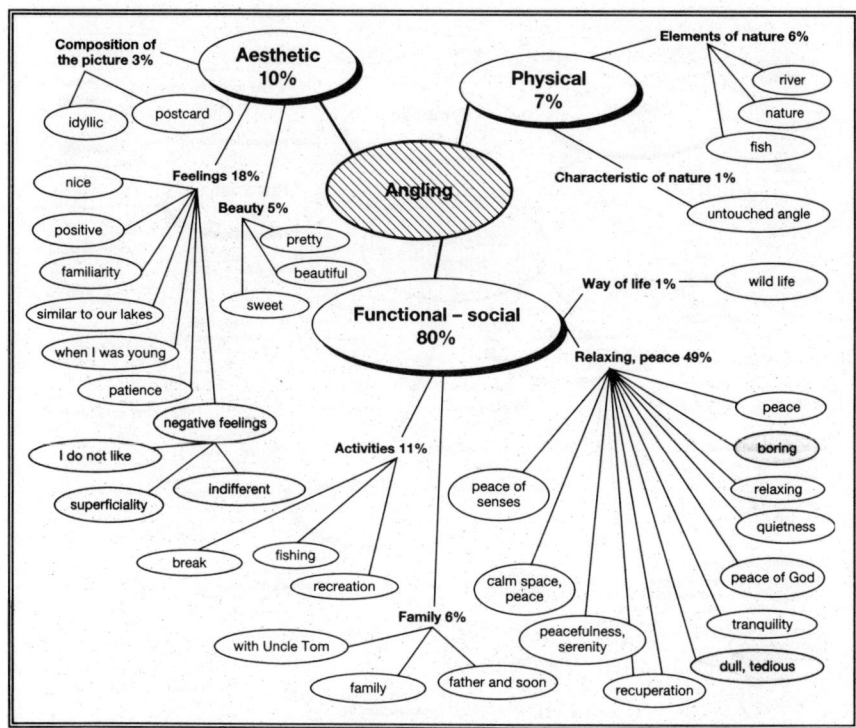

Figure 6.4 Italian respondents' comments on picture 'Angling'

Finns or those familiar with the cultural particularities. Without any introduction to Finnish culture, these meanings remain rather irrational and even artificial in the mental images of the non-Finnish audience (see Tuohino & Pitkänen 2004b; Tuohino et al. 2004).

The respondents approached the pictures both as outside observers and as participating actors. The transformation of a neutral lake landscape into a meaningful experience depended on the mental images evoked by both the aesthetic content of the picture and its internal elements. The familiarity of the landscape acted as a catalyst for the topophilic sensations of the respondents. Correspondingly, lack of familiarity often aroused negative feelings. However, interest in a new place and over-familiarity with a well-known place should also be taken into account.

A number of similarities can be observed when looking at the categorical distribution of picture-specific mental images across different groups of respondents. A few pictures contained unfamiliar elements to certain respondent groups, which turned them from participants into observers of the landscape. Similarly, a trade-off can be observed between the 'aesthetic' and 'functional-social' images. It would seem that if the

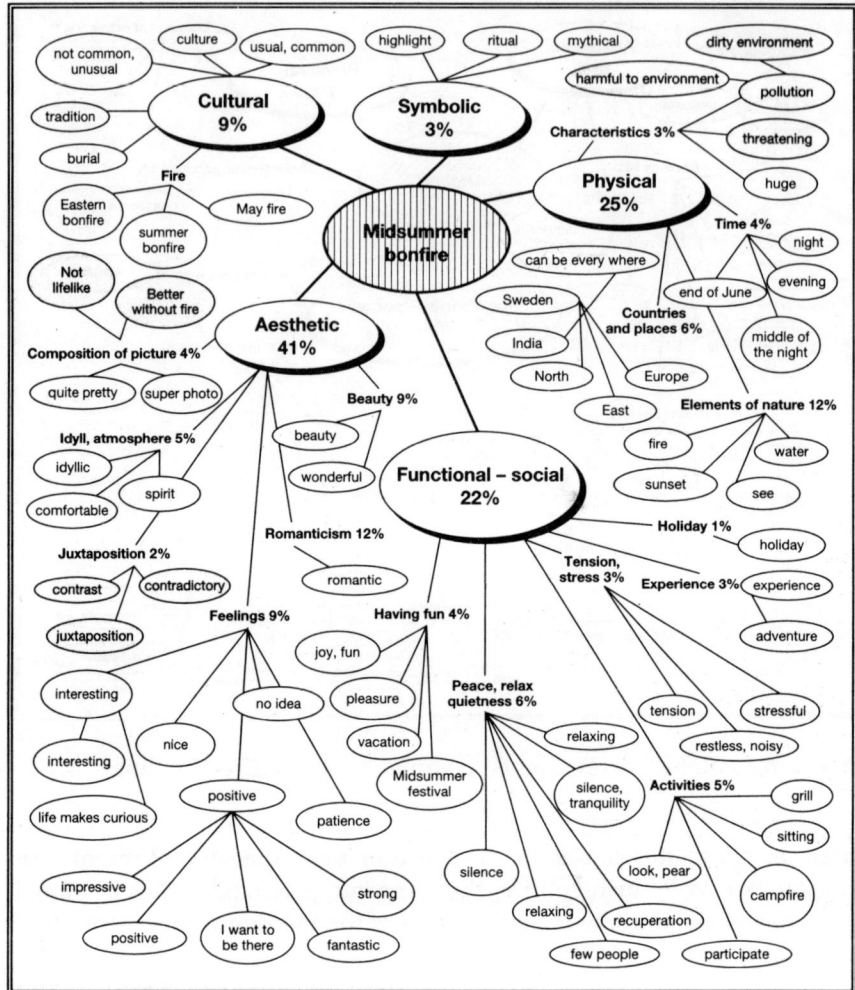

Figure 6.5 German respondents' comments on picture 'Midsummer Bonfire'

contents of a picture do not give functional or social identification points, they are sought in the aesthetic appearance of the picture or even in cultural cues. The differences between countries in the way the images were distributed were rather small. Relatively more German than Italian responses were coded into the categories 'functional' and 'physical', while more Italian than German responses fell into the category 'aesthetic' – i.e. mental images connected with the appearance of the picture and, especially, personal feelings that were evoked.

In Search of the Spirit of the Lake

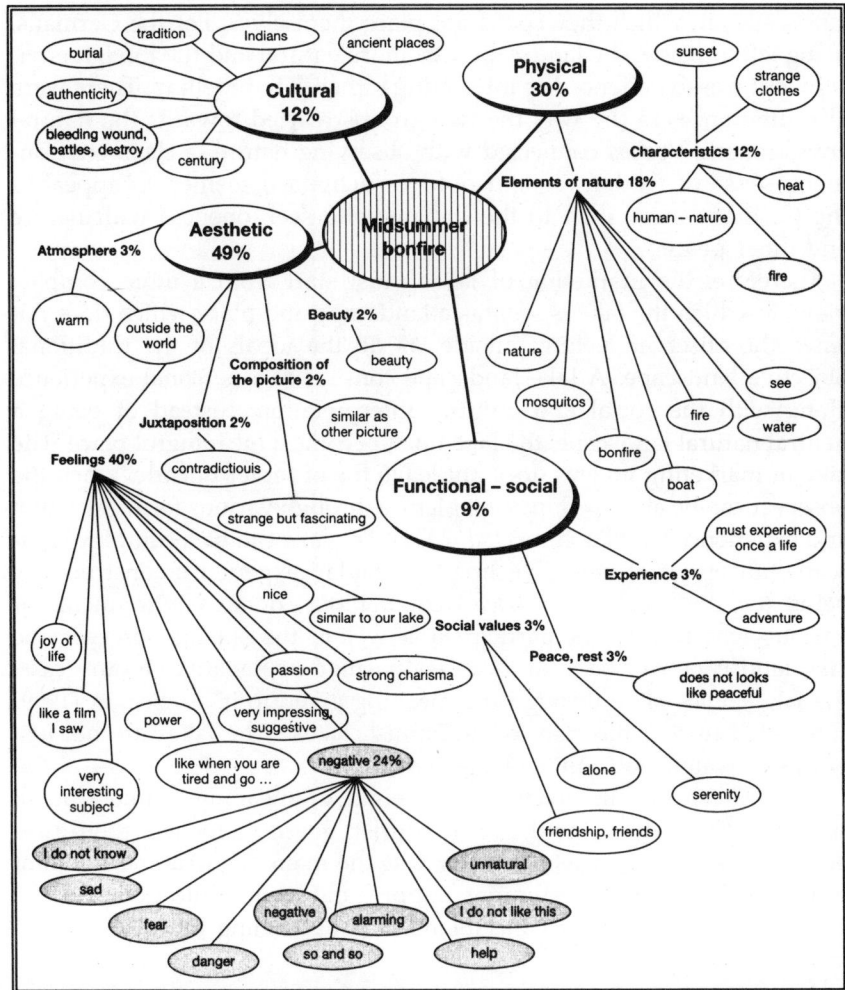

Figure 6.6 Italian respondents' comments on picture 'Midsummer Bonfire'

When we look at the proportion of negative reactions rather than different categories of reactions, differences between countries are more marked. The reactions of the Germans were most negative towards the picture about frolicking in the water, which was considered as too lighthearted and amusing. The Italians reacted most negatively towards 'Midsummer Bonfire'. The picture of a bonfire raised concern, even fear, in people who are ignorant of this summer tradition. In comparison with the Germans, the Italians reacted negatively towards pictures showing landscapes in which they could not place themselves. For the Germans, being left 'outside' the landscape was more natural and the foreign environment was experienced as interesting rather than repelling. There were also differences in the way the two groups reacted towards the natural environment. Values connected with observing nature (a clean environment, sense of wilderness and outdoor activities) seemed to appeal to the Germans more than to the Italians who seek oneness with nature and inner peace.

Therefore, the marketing of lakes must start from a more complex view, in which the lake is seen as a landscape and place which does not leave the observer feeling outside, as do the ideals of the traditional objective landscape. A lake landscape consists of a personal experience along with the social and cultural interpretation. Instead of being a neutral natural landscape, the lake must become a meaningful place. The lake of marketing images does not leave the observer outside, when the observer can be absorbed into the picture through various forms of action and experience (see Tuohino et al. 2004). The lake can be a complex landscape, an object of strong feelings, a holiday experience construed by being together with one's travel companions, or an environment for various activities. To the participant as well as the outside observer, the lake landscape can be seen as either positive or negative. In any case, the lake comes alive through the meaning given to it. As Seaton (1999) concluded during his visit to the Finnish lakes: 'The lakes are not just ponds of water, but ponds of meanings.'

The waterscape, in general, as a pond of meanings came true in summer 2004 when the Finnish Tourist Board launched their new Summer Campaign in which water was the main element and the joint principal factor of the summer supply and communication. As the campaign stated, 'Summer in Finland is A Refreshing Journey'.

References

Crompton, J.L. (1979) Motivations for pleasure vacation. *Annals of Tourism Research* October/December VI (4), 408–24.

Echtner, C. and Ritchie, J.R.B. (1993) The measurement of destination image: An empirical assessment. *Journal of Travel Research* 31 (4), 3–13.

Eyles, J. (1985) *Senses of Place*. Warrington: Silverbrook.

Haarni, T. and Karvinen, M. and Koskela, H. and Tani, S. (1997) Johdatus nyky-maantieteeseen (Introduction to today's geography). In T. Haarni, M. Karvinen, H. Koskela and S. Tani (eds). *Tila, paikka ja maisema. Tutkimusretkiä uuteen maantieteeseen* (Space, Place and Landscape. Excursions into New Geography) (pp. 9–34). Tampere: Vastapaino.

Hall, C.M. and Page, S.J. (1999) *The Geography of Tourism and Recreation. Environment, Place and Space*. London and New York: Routledge.

Hemmi, J. and Vuoristo, K-V. (1993) *Matkailu* (Tourism). Porvoo: Profit. WSOY.

Hudman, L.E. and Jackson, R.H. (1990) *Geography of Travel and Tourism*. New York: Delmar Publishers.

Järviluoma, J. (1994) Matkailun työntö- ja vetovoimatekijät ja niiden heijastuminen lomakohteen valintaan. (Push and pull factors in tourism and their reflections on choosing of holiday destination) In S. Aho (ed.) *Matkailun vetovoimatekijät tutkimuskohteina* (Tourism Attractions as Research Objects) (pp. 31–48). Oulu: University of Oulu, Research Institute of Northern Finland.

Kanninen, V. (1993) Maiseman lukeminen: kartografia, valokuvaus ja kirjallisuus mielikuvien muokkaajina (Interpretation of Landscape: Cartography, Photography and Literature as Revisers of Images) In P.T. Karjalainen (ed.) *Maantieteen maisemissa. Aiheita kulttuurimaantieteen alalta* (In the Landscape of Geography. Essays on Cultural Geography) (pp. 19–28). Bulletin No. 25. Joensuu: University of Joensuu. Human and Planning Geography.

Karjalainen, P.T. (1986) *Geodiversity as a Lived World: On the Geography of Existence*. Joensuu: Publications of Social Sciences No. 7, University of Joensuu.

Karjalainen, P.T. (1995) Mahdollisten maisemien sematiikkaa (Sematic of Landscapes of Possibilities). *Terra* 107 (2), 123–5.

Koskela, A. (2002) Veden valta. Järvi tunteiden kokemisen paikkana (Power of Water. Lake as a Place of Experiencing Feelings). *Muuttuva matkailu – tietoa matkailusta ja matkailuelinkeinosta (Tourism in Transition – Information on Tourism and Tourism Industry)* 2/2002, 74–6.

Lehtinen, A. (2001) Järvimaiseman lukeminen (Interpreting the Lake Landscape). Unpublished presentation, Spirit of the Lake Symposium on 22 November, Valamo Monastry, Heinävesi, Finland.

Meinig, D.W. (1979) The beholding eye. Ten versions of the same scene. In D.W. Meinig (ed.) *The Interpretation of Ordinary Landscapes* (pp. 33–48). Oxford: Geographical Essays, Oxford University Press.

Meyer, J.L. (2001) Nature preservation, sense of place and sustainable tourism: Can the 'Yellowstone experience' survive? In S.F. McCool and R.N. Moiley (eds) *Tourism, Recreation and Sustainability. Linking Culture and the Environment* (pp. 91–104). Oxford: CABI Publishing.

Middleton, V.T.C with Clarke J. (2001) *Marketing in Travel and Tourism*. 3rd edn. Oxford: Butterworth Heinemann.

Middleton, V.T.C. (1997) Sustainable tourism: A marketing perspective. In M.J. Stabler (ed.) *Tourism and Sustainability. Principles to Practice* (pp. 129–42). New York: CAB International.

Moisey, R.N. and McCool, S.F. (2001) Sustainable tourism in the 21st century. In S.F. McCool and R.N. Moiley (eds) *Tourism, Recreation and Sustainability. Linking Culture and the Environment* (pp. 343–52). New York: CABI Publishing.

Paasi, A. (1986) *Neljä maakuntaa. Maantieteellinen tutkimus aluetietoisuuden kehittymisestä* (Four Provinces. Geographical Research on Development of Conciousness of Region). Joensuu: Publications of Social Sciences No. 8. University of Joensuu.

Park, C.C. (1994) *Sacred Worlds. An Introduction to Geography and Religion*. London: Routledge.

Pitkänen, K. and Vepsäläinen, M. (2003) Sinisen Saimaan Savonlinna. Vesistöjen merkitys Savonlinnan matkailuimagolle (Savonlinna, The Pearl of Lake Saimaa. The Importance of Lakes in the Tourist Image of Savonlinna). *Alue ja Ympäristö* 32 (1), 33–45.

Raivo, P. (1995a) Kulttuurimaiseman käsitteellistyminen maantieteellisessä tutkimuksessa. (Conceptualisation of Cultural Landscape in Geographical Research). In P.T. Karjalainen and P.J. Raivo (eds) *Johdatusta kulttuurimaantieteelliseen maisematutkimukseen* (Introduction to Humanistic Landscape Research) (pp. 7–20). Teaching Hand-out No 21. Oulu: University of Oulu, Department of Geography.

Raivo, P. (1995b) The discursive transformation of the religious landscape: The case of the Orthodox Church in Finland. *Nordia Geographical Publications* 24 (2), 39–52.
Raivo, P. (1997) Kulttuurimaisema: alue, paikka vai tapa nähdä (Cultural Landscape: Region, Place or a Way of Seeing). In T. Haarni, M. Karvinen, H. Koskela and S. Tani (eds) *Tila, paikka ja maisema. Tutkimusretkiä uuteen maantieteeseen* (Space, Place and Landscape. Excursions into New Geography) (pp. 193–209). Tampere: Vastapaino.
Relph, E. (1976) *Place and Placelessness*. London: Pion.
Saarinen, J. (1995) Matkailualueen hahmottuminen ja matkailun vetovoimatekijät: esimerkkinä Saariselän matkailualue (Formation of Tourism Region and Tourism Attractions; Case Saariselkä Tourism Region). In S. Aho and H. Ilola (eds) *Matkailu alueellisena ilmiönä* (Tourism as Regional Phenomenon) (pp. 105–22). Oulu: University of Oulu, Research Institute of Northern Finland.
Seaton, A. (1999) Järvi ei ole vain vesiallas. Interview in Ita=Savo 21.9.1999.
Sepänmaa, Y. (1998) Matkan estetiikkaa. Matka koettuna ja kokemus teoksena (Aesthetic of Travel. Travel as Experience and Experience as Work). In M-L. Hakkarainen and T. Koistinen (eds) *Matkakirja. Artikkeleita kirjallisista matkoista mieleen ja maailmaan* (Travel Book. Articles of Written Trips to Mind and World) (pp. 13–23). Research Reports of Literature and Culture, University of Joensuu, Faculty of Humanities.
Shamai, S. (1991) Sense of place: An empirical measurement. *Geoforum* 22 (3), 347–58.
Suvantola, J. (2002) *Tourist's Experience of Place*. Burlington: Ashgate.
Tarasti, E. (1990) Johdatusta semiotiikkaan. Esseitä taiteen ja kulttuurin merkkijärjestelmistä (Introduction to Semiotics. Essays of Signs of Art and Culture). Gaudeamus, Helsinki.
Tuan, Y-F. (1974) *Topophilia. A Study of Environmental Perception, Attitudes and Values*. Englewood Cliffs, NJ: Prentice-Hall.
Tuan, Y-F. (1977) *Space and Place. The Perspective of Experience*. Minneapolis, MN: University of Minnesota Press.
Tuan, Y-F. (1979). Thought and landscape. The eye and the mind's eye. In D.W. Meinig (ed.) *The Interpretation of Ordinary Landscapes* (pp. 89–102). Oxford: Geographical Essays, Oxford University Press.
Tuohino, A. and Peltonen, A., Aho, S., Eriksson, R., Komppula, R. and Pitkänen, K. (2004) *The Image of Finland in the Seven Main Market Areas*. MEK. A: 140. Helsinki: Finnish Tourist Board.
Tuohino, A. and Pitkänen, K. (2004a) Selling waterscapes? In J. Saarinen and C.M. Hall (eds) *Nature-Based Tourism Research in Finland: Local Contexts, Global Issues* (pp. 129–50). Finnish Forest Institute, Research Papers 916.
Tuohino, A. and Pitkänen, K. (2004b) The transformation of a neutral lake landscape into a meaningful experience – interpreting touristic photos. *Journal of Tourism and Cultural Change* 2 (2), 77–93.
Tuohino, A. and Uusi-Illikainen, M. (1997) Kulttuurimatkailu Ateenassa ja Roomassa (Cultural Tourism in Athens and Rome). Unpublished Master's Thesis, Department of Geography, University of Oulu.
Walsh, J.A., Jamrozy, U. and Burr, S.W. (2001) Sense of place as a component of sustainable tourism marketing. In S.F. McCool and R.N. Moiley (eds) *Tourism, Recreation and Sustainability. Linking Culture and the Environment* (pp. 195–216). New York: CABI Publishing.
Vuoristo, K-V. and N. Vesterinen (2001) *Lumen ja suven maa. Suomen matkailumaantiede* (Land of Snow and Summer. Tourism Geography of Finland). Helsinki: WSOY.

Chapter 7
Lake Tourism in the Netherlands

MARTIN GOOSSEN

Introduction

The Netherlands is a relatively small country in the delta of the rivers Rhine, Maas and Schelde in north-western Europe. The name Netherlands means 'low-lying land'. About half of the Netherlands is below sea-level and 18% of its surface area is covered by water. The Dutch have a special relationship with water. Over the years, it has been a constant struggle to keep the water out, but it is also an integral element of the society, culture and environment. Water is so important that three ministries are involved in developing lake tourism, although this does make policy and planning very complex. The Ministry of Transport, Public Works and Water Management is responsible for mobility policy (also for water) in the Netherlands, and for protection against floods or falling water tables. The Ministry of Economic Affairs is responsible for tourism. The Ministry of Agriculture, Nature and Food quality is responsible for recreation. A great deal of recreation takes place on or by water and many tourists visit the Netherlands to enjoy activities on the water. To make it even more complex, some provinces are also responsible for maintaining lakes and canals/rivers.

Because of the enormous supply of water in the Netherlands, there are many of opportunities for lake tourism activities such as swimming, fishing, sailing, canoeing and ice-skating. The following types of water areas can be distinguished:

- lakes and pools;
- canals and waterways;
- rivers;
- large inland water;
- the North Sea.

There are 75 lakes larger than 217 acres (100 ha) and there is a total 3728 miles (6000 km) of waterways. Of this total, 2734 miles (4400 km) are officially designated by the government as primary recreational waterways.

Trade shipping uses around 994 miles (1600 km) of this network. The biggest lakes or inland water are the IJsselmeer (4,80,000 ha) and the Waddenzee (140,000 ha) with connections to the German and Danish Wadden. The Netherlands has three important lake districts: The Delta (21,000 acres (85,000 ha)), Friesland lake district (24,710 acres (10,000 ha)) and the Holland lake district (160,618 acres (6,500 ha)). A fourth is being developed – namely the Maas lake district (7,413 acres (3,000 ha)) (Figure 7.1). Because of the nature and size of the various water areas, each location has specific uses and therefore different types of watercraft. Important elements that influence their use are: the size and depth of the water, the presence of streams or tidal streams, the presence of locks and the presence of low bridges.

Although lakes are an important component of coastal and inland waterways, it is difficult to isolate the contribution of lakes from other

Figure 7.1 Coastal and inland waterways of the Netherlands

aspects of marine tourism. Therefore, in identifying the characteristics and activities of lake users, we will examine the contribution of 'water-related tourism', which is defined as tourism in which the core attraction is water, rather than lake tourism per se. Water-related tourism is of great economic significance in the Netherlands, with total expenditure estimated by the Netherlands Research Institute Recreation and Tourism (NRIT)/Government Institute for Coast and Sea (RIKZ) (2002) at approximately €2.5 billion a year and €4 billion by the Stichting Recreatie (2002). The different estimates of economic impact are also reflected in estimates of the number of jobs generated by water-related tourism, with numbers ranging from 32,000 (NRIT/RIKZ) to 92,000 jobs (Stichting Recreatie 2002) (Table 7.1).

Since 1992 pleasure craft have been counted at several locks. These counts provide an insight into which lock (or the area in which the lock is situated) is the most popular for recreational boating. Figure 7.2 illustrates that the total number of lock passages has stayed at around 700,000 craft a year (Goossen & Langers 2002). However, because watercraft can be counted more than once on the same trip, the figure does not indicate if the total number of craft has remained the same, as it is possible that there are fewer craft making longer trips, or maybe more craft that are making shorter trips (Goossen & Langers 2002).

Comparisons Between 1993 and 2002 Surveys

Because water-related tourism is important in the Netherlands, the government is interested to know more about water-based tourists, especially those who make a trip of several days. Therefore, research on

Table 7.1 Summary of water-related tourism in the Netherlands

Number of open-air swimming locations	600
Number of marinas	1,525
Number of permanent berths	150,000
Number of mooring facilities[1]	84,500
Number of motor craft[1]	126,000
Number of sailing boats[1]	126,000
Number of canoes[1]	135,000
Estimated annual expenditure	€2.5 billion[1], €4 billion[2]
Estimated employment generated	32,000[1], 92.000[2]

Sources: 1 Netherlands Research Institute Recreation and Tourism (NRIT)/Government Institute for Coast and Sea (RIKZ) 2002; 2 – Stichting Recreatie 2002

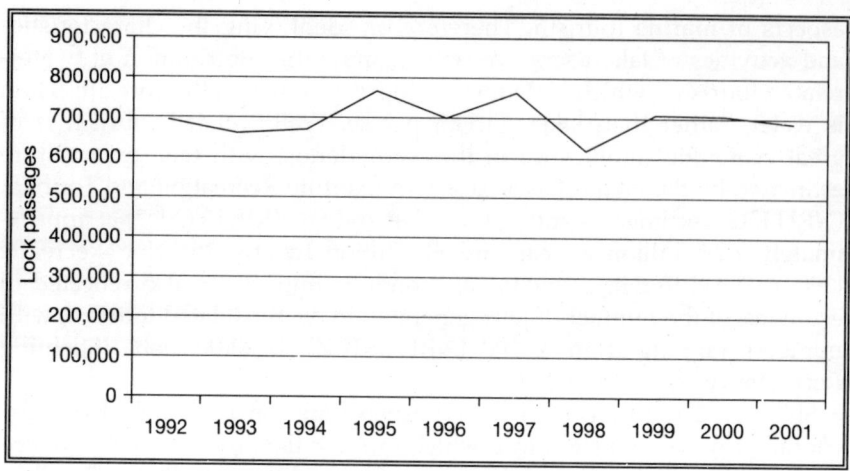

Figure 7.2 Number of passages of pleasure craft counted in locks

water-related tourism in the summer of 2002 (Goossen & Langers 2002) that replicated research that was undertaken in 1993 (de Bruin & Klinkers 1994) in order to provide comparative data.

Because of the different lakes, rivers and canals and their specific watercraft, the research was undertaken throughout the Netherlands, in order to take into account the specific characteristics of individual water areas. The research carried out done in 13 area, 5 visited mainly by sailing boats (lakes and inland waters) and 6 areas visited mainly by motor craft (primarily rivers and canals). Two areas are visited by both sailing boats and motor craft. In total, 1454 persons were interviewed – 671 skippers of sailing boats and 783 skippers of motor craft. In the face-to-face interviews questions were asked about the characteristics of the boat, and the motivations and intentions of the users and their holiday trip.

The Skippers

Table 7.2 indicates that skippers are getting older. In 1993 the mean age was 45 years and in 2002 the mean age of skippers was about 51 years. There was an increase particularly in the percentage age between 50 and 64 years. Table 7.2 also indicates that skippers of motor craft are older than skippers of sailing boats. This result is consistent with the trend of the increase in the average age of the Dutch population. Table 7.3 shows the number of people on board. The average number for motor craft is 2.6 persons and for sailing boats is 2.9 persons. This number decreased in 2002 compared to 1993. There is a strong relationship with the average age.

Lake Tourism in the Netherlands

Table 7.2 Age (%) of skippers per type of boat, 1993 and 2002

Age in years	Motor crafts		Sailing boats	
	2002	1993	2002	1993
15–29	2	5	8	15
30–39	11	16	15	20
40–49	19	32	30	42
50–64	51	36	40	22
≥ 65	17	10	7	2
Mean	55	48	48	42

Table 7.3 Number of persons on board (%) per type of boat, 1993 and 2002

Number of persons	Motor crafts		Sailing boats	
	2002	1993	2002	1993
1–2	69	54	56	47
3	10	17	16	15
4	15	21	20	29
≥5	6	7	8	10
Mean	2.6	2.8	2.9	3.0

Most of the skippers in the survey were Dutch, but about 10% of respondents were foreign. Of these, 75% were German, 13% Belgian, 5% British, 7% other nationalities. Of these foreign skippers, about 65% had their boat based at a marina in the Netherlands.

The boats

The economy has grown in the last ten years in the Netherlands and this is reflected in the type and size of boats that are being purchased. The average boat is longer, wider, higher, has more depth and is more luxurious than in 1993 (Table 7.4). Most of the boats have navigational support equipment, including sailing maps, fish-finders and Global Positioning Systems.

Table 7.4 Dimensions per type of boat, 1993 and 2002

Mean	Motor crafts		Sailing boats	
	2002	1993	2002	1993
Length	10.2	9.7	9.4	8.9
Width	3.3	3.1	3.1	2.9
Height	2.6	2.2	12.4	11.9
Depth	1.0	0.9	1.3	1.2

The routes

The respondents were able to indicate on a map the route they had already sailed and the route they were planning to sail. Also, they could indicate the starting point of the total trip (generally the marina or home port), the planned destination(s), the starting point of the day of interview and the final destination of the route to be sailed on the day of the interview. All their routes were digitalised (see Figures 7.3 and 7.4 for the results).

There is a remarkable difference in the routes taken by the motor craft and the sailing boats. Most motor craft use a river or canal, while sailing boats have a preference for lakes and large inland water. However, as the figures indicate, almost every lake or waterway has been used. Another remarkable result, taking into account the relatively short distances between lake districts, is that sailors tend to stay in the region closest to their home port. So, people with a boat in the Friesland lake district sail in that district and the area close to it. Sailors in the Delta region stay there or in the region close to it. The centre of the Netherlands is still the most popular area. There is also a difference in skippers' choice of boating areas. Skippers of sailing boats tend to choose more often the region where they always sail. They select a well-known area and have no need to discover new regions. The skippers of motor craft, however, are the opposite. In the last nine years they tour more and more, venturing further afield and discovering new regions.

Approximately 40% of the skippers also say that occasionally they sail abroad, so sailing is becoming more European. Dutch sailors go abroad, and German and Belgian sailors visit the Netherlands. This trend will probably increase over time because of the characteristics of the skippers (older with more free time and more money to spend).

Lake Tourism in the Netherlands

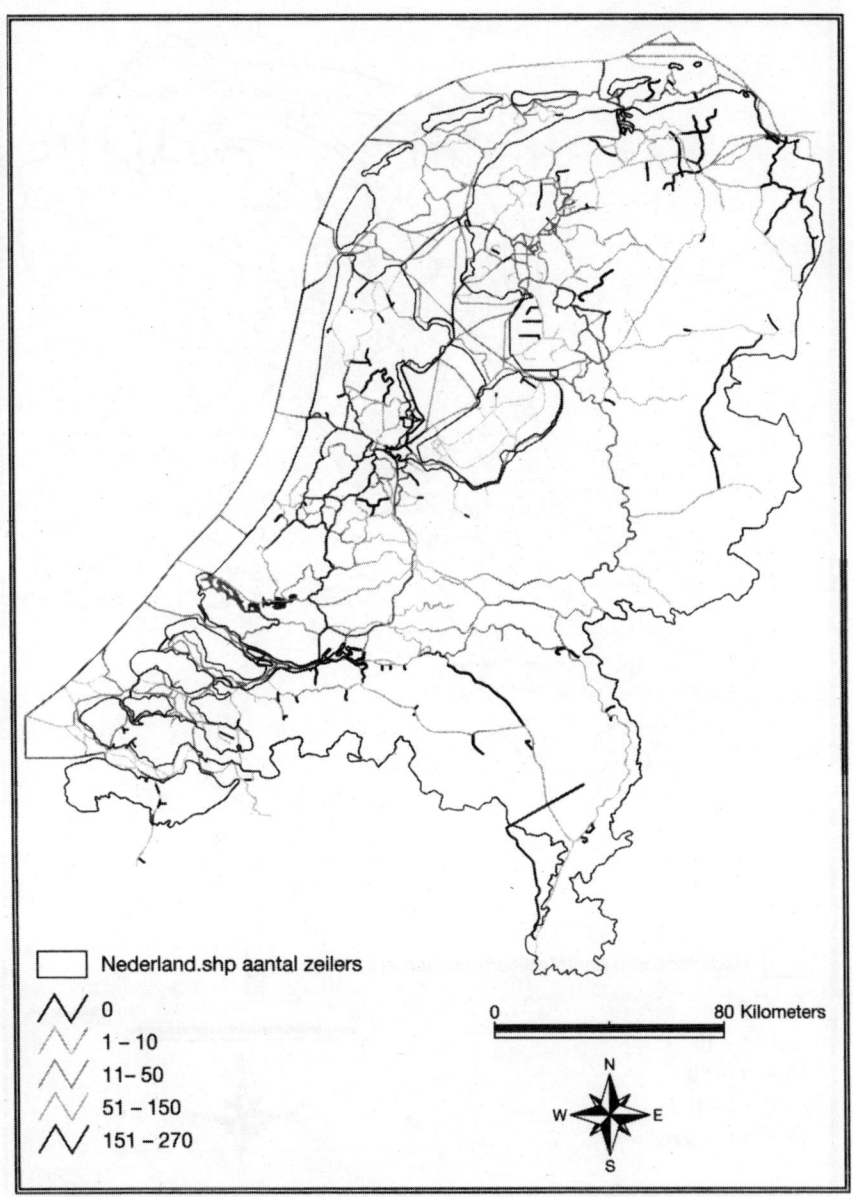

Figure 7.3 Use of the waters by sailing boats

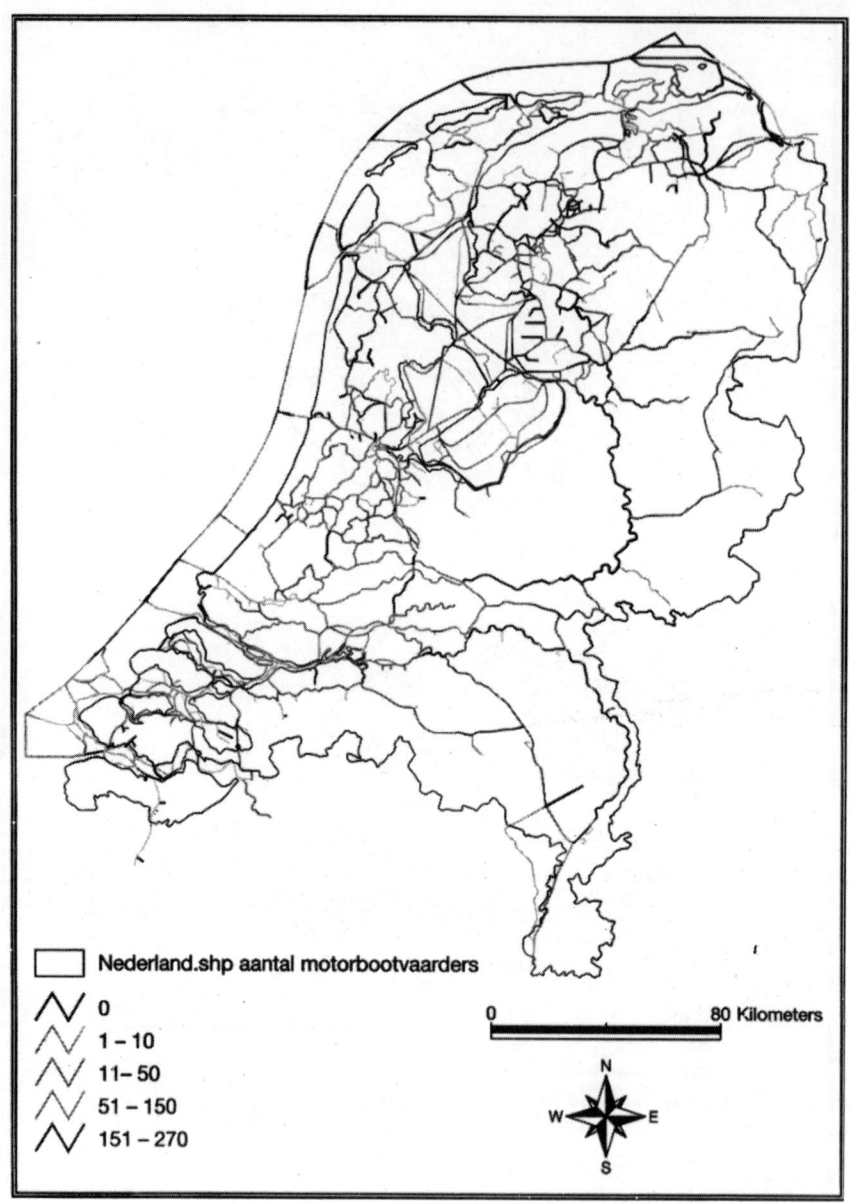

Figure 7.4 Use of the waters by motor craft

The most popular reasons why skippers choose a region for boating activities are:

- diversity of water types;
- old cities and villages;
- nature and landscape.

These reasons apply to almost every region. However, the respondents also gave other reasons for their choice of an area. Some regions are popular because there are opportunities for cycling and hiking, others for their tides or the amount of landing stages. Therefore, water-related tourism is not just about sailing. It is a combination of sailing and what 'the land' has to offer. The facilities on the boundaries between regions are also important. Indeed, one of the results of this research is that the tourist information boards are increasingly promoting the land opportunities for recreation on land surrounding lake districts, as well as the facilities available.

The most popular reasons why skippers choose a specific day route are:

- location of old cities and villages;
- shortest route to destination;
- the weather;
- attractive landscape.

Sailors also choose a specific route because it is familiar. Skippers of motor craft often choose a particular route so that they can stop and do some shopping.

The Use of the Boat and Expenditure

About 66% of skippers indicated that during the course of a year they spend on average six days on board simply relaxing, or repairing/maintaining their boat. Skippers of motor crafts spend more time on board without making a trip than sailors (Table 7.5). Skippers use their boats about 65 days a year. On average, sailors make a holiday trip of 21 days, while skippers of motor craft make a holiday trip of 33 days on average.

During this time sailors spend €72 per day per boat and skippers of motor craft spend on average €77 per day per boat. In total, the water-related tourism aggregate is an average of €75 each day per boat. Table 7.6 shows the amount of expenditure on different items. Shopping and visiting restaurants and cafes are the items with the highest daily expenditure.

Table 7.5 Median use of boat, per type of boat

Use	Sailing boat	Motor craft	Total
On board, not sailing	5	10	6
Daytrip	6	4	5
Weekend trip (1 overnight)	10	8	10
Holiday (2 or more overnights)	25	30	28
Total amount of days	56	74	65

Source: Goossen and Langers (2002)

Table 7.6 Mean expenditure (€) of respondents during last 24 hours per type of boat

Expenditure	Sailing boats €	Motor craft €
Daily shopping	22.00	23.90
Visit restaurant/café	20.51	19.34
Recreational shopping	8.29	10.14
Fuel and shipping items	5.43	12.86
Overnight berth fees	8.40	5.63
Other expenditures	4.84	2.68
Visit towns, villages, attractions, museums	1.22	1.22
Berth fees during the day	0.56	1.17
Bridge and lock fees	0.68	0.55
Total	71.93	77.49

Source: Goossen and Langers (2002)

The Problems

The main problems encountered during a holiday trip are given in Table 7.7. The behaviour of inexperienced water-sports people is regarded as the biggest problem. The introduction of a sailing licence is under consideration, but such an initiative is being opposed by water-sports organisations because they believe that it could diminish the tourist attractions of the Netherlands for boat enthusiasts.

Table 7.7 Percentage of respondents reporting encounters with specific problems during a boating holiday

Problem	Total percentage of respondents
Behaviour of inexperienced water-sports persons	49
Number of landing stages at bridges and locks	40
Service times at bridges and locks	35
Too few sanitary facilities at berths	28
Areas with fixed bridges with too low headroom	25
Too many other water-sports people on the water	24
Areas with insufficient water depth	19
Berth facilities in rural areas	19
Too few facilities for disposing of refuse/glass/oil and emptying WCs	18
Berth facilities in urban areas	18
Water pollution	14
Lack of up-to-date tourist (sailing) information	12
Problems with trade shipping	8
Water plants	7
Connections to attractions/recreational facilities on land	6

Source: Goossen and Langers (2002)

Product Development and Marketing

A specific organisation has been established to solve problems with landing stages, bridges and connections between waterways. The national government, provinces and water-sports organisations have joined forces and agreed to establish a co-operative project aimed at solving the problem of bottlenecks in the Dutch network of waterways. These bottlenecks hamper recreational sailing and it is hoped that the problem will be solved by 2012. In 1995, the Dutch Recreational Waterways Foundation (DRW) was created (see www.srn.nl). The government and provinces decided to support the DRW for 20 years, with a contribution of approximately €2.5 million per year. The DRW is acting as a co-financier in order to stimulate the provinces, municipalities and waterboards into renovating the waterways under their control. For example, some former bridges have now been replaced by aquaducts. The DRW is trying to fulfil the wishes of boat and yacht enthusiasts, and to make the Netherlands even more attractive for water-related tourism. However, the DRW is developing the product but has no specific marketing goals. Instead, the regional and local tourist-information boards are responsible of the marketing of water-related tourism. In addition, they also have competitors, such as the Dutch ANWB (an automobile club like the ADAC in Germany, and the AA and RAC in the UK), that also play an important role in the promotion of water-related tourism through the production of maps of the waterways, brochures and new route development.

The Netherlands Board of Tourism and Conventions (part of the Ministry of Economic Affairs) aims to promote tourism and leisure to and within Holland. Every year, the Netherlands Board of Tourism and Conventions (NBTC) selects a theme to attract more tourists. The idea behind a central theme is to combine all marketing efforts – for example, one theme in the past was Vincent Van Gogh. However, it is a testimony to the importance of water-related tourism that in 2005 the central theme was 'Water'.

References

De Bruin, A.H. de and Klinkers, P.M.A. (1994) *Recreatietoervaart; de moeite waard*. Rapport 307. Wageningen: Staring Centrum.

Goossen, C.M. and Langers, F. (2002) *Recreatietoervaart; 9 jaar later*. Rapport 627. Wageningen: Alterra, Green World Research.

Netherlands Research Institute Recreation and Tourism (NRIT)/Government Institute for Coast and Sea (RIKZ) (2002) *De betekenis van water voor recreatie en toerisme in Nederland* (The Meaning of Water for Recreation and Tourism in the Netherlands). Den Haag: Netherlands Research Institute Recreation and Tourism/Government Institute for Coast and Sea.

Stichting Recreatie (2002) *Cijfermateriaal boven water. Inventarisatie feiten en cijfers waterrecreatie* (Data Pop-Up: Facts and Data of Water Recreation). Den Haag: Stichting Recreatie.

Chapter 8
The Ostroda-Elblag Canal in Poland: the Past and Future for Water Tourism

GRAZYNA FURGALA-SELEZNIOW, KONRAD TURKOWSKI, ANDRZEJ NOWAK, ANDRZEJ SKRZYPCZAK AND ANDRZEJ MAMCARZ

Introduction

Taking part in tourist and recreational activities is a way of spending leisure time. Tourist migrations are conditioned by the natural and cultural assets of the geographical environment. Areas that are relatively less populated and industralised, characterised by diverse landscapes rich in forests and lakes, can play a particularly important role in tourism development. Such is the region containing a historic water route of lakes connected by the Ostroda–Elblag Canal (also known as the Elblag Canal) in Poland.

Since the Middle Ages the only waterway that connected the environs of Ostroda and Ilawa with the Vistula river and the Baltic Sea was the Drweca river (Januszewski 2001). However, a waterway that would travel via Elblag would shorten this important medieval trading route nearly fivefold. Therefore, a decision to build a waterway from east Prussia to the Baltic Sea was made in the 1820s by the Council of the Prussian province, although a decision to allocate financial resources to the construction of the canal was not made by the Prussian King, Frederic Wilhelm III, until 1837 (Januszewski 2001). The construction of the canal commenced in 1844. The canal's constructor, the engineer Georg Jacob Steenke, worked out a feasible project to overcome land obstacles between Lakes Drweckie, Ilinsk, Ruda Woda, Sambor, Piniewo and Druzno. The biggest difficulty was the difference between the water levels in the various lakes which reached about 109 ys (100 m) over a distance of 6 miles (9.6 km). It was therefore necessary to build locks at Zielona and Milomlyn, accompanied by a system of slipways that would enable ships to traverse land elevations. To achieve this, Steenke drew on the canal engineering solutions used near Ketley in England and in the Morris's Canal in Pennsylvania, USA.

The first ships from Ilawa to Milomlyn set out in October 1860 (Baldowski 1998); the entire canal was fully operational in 1872. Until 1912, the canal was used exclusively for the transportation of agricultural and industrial products. Shortly before the First World War, however, it began to serve tourism and recreational functions, and soon became a tourist attraction. Regular summer tourist cruises began in 1927. In 1978 the canal was recognised as a nationally designated monument of technology by the Polish government and since then has been registered as a historic object of unique value in the register of the historic objects in the *voivodship* (province) of Warmia and Mazury.

Today the canal primarily serves tourist and recreational functions. By combining thousands of hectares of lake waters into water routes, that run across highly diverse and well-forested terrains, the canal creates a system that allows visitors to pursue many types of tourism and water-based activities (e.g. travelling by ship, sailing, canoeing, angling, sunbathing) that makes it a significant tourist attraction and recreational resource.

The Canal Route

According to the geographical division of Poland, the route of the canal runs across the geographically designated macroregion known as the Pojezierze Wschodniopomorskie (Wschodniopomorskie Lake District, north Poland), which includes the mesoregions of Zulawy Wislane (a depression in the delta of the Vistula) and Pojezierze Ilawskie (Ilawskie Lake District) (Kondracki & Ostrowski, 1973–8). The canal is a branched navigation system with a total length of 116 miles (187.2 km), which links over 15 lakes and navigable stretches of rivers flowing to Lake Druzno and Zalew Wislany (the Vistula estuary) (Baldowski 1998). Four chamber locks help to harness the differences in water levels between the lakes. Two locks are situated along the Ostroda–Stare Jablonki route, at Mala Rus and Ostroda (Figure 8.1). Two other locks occur along the Ostroda–Elblag route, at Zielona and Milomlyn. Five slipways (at Buczyniec, Katy, Olesnica, Jelenie and Caluny) built between the lakes north of Milomlyn and Lake Druzno allow ships to travel over land elevations. They are located over a distance of 6 miles (9.6 km) with a 109 yd (99.5 m) decline.

The slipway operates like a ski lift, 'carrying' ships through the canal (Bakowska, 1996). Ships are tied to side rails on submerged platforms, which are 22 yd (20 m) long and can carry up to 20 tons. A steel rope passing through gear wheels connects the platform with another platform on the ramp. When the platform with the ship is raised up the ramp, the other platform descends; they pass each other midway. At the other end of the slipway, the platform is submerged again and the ship sails on. Power for the slipway machinery is supplied by water, which moves waterwheels and a system of pulleys.

The Ostrada–Elblag Canal, Poland 133

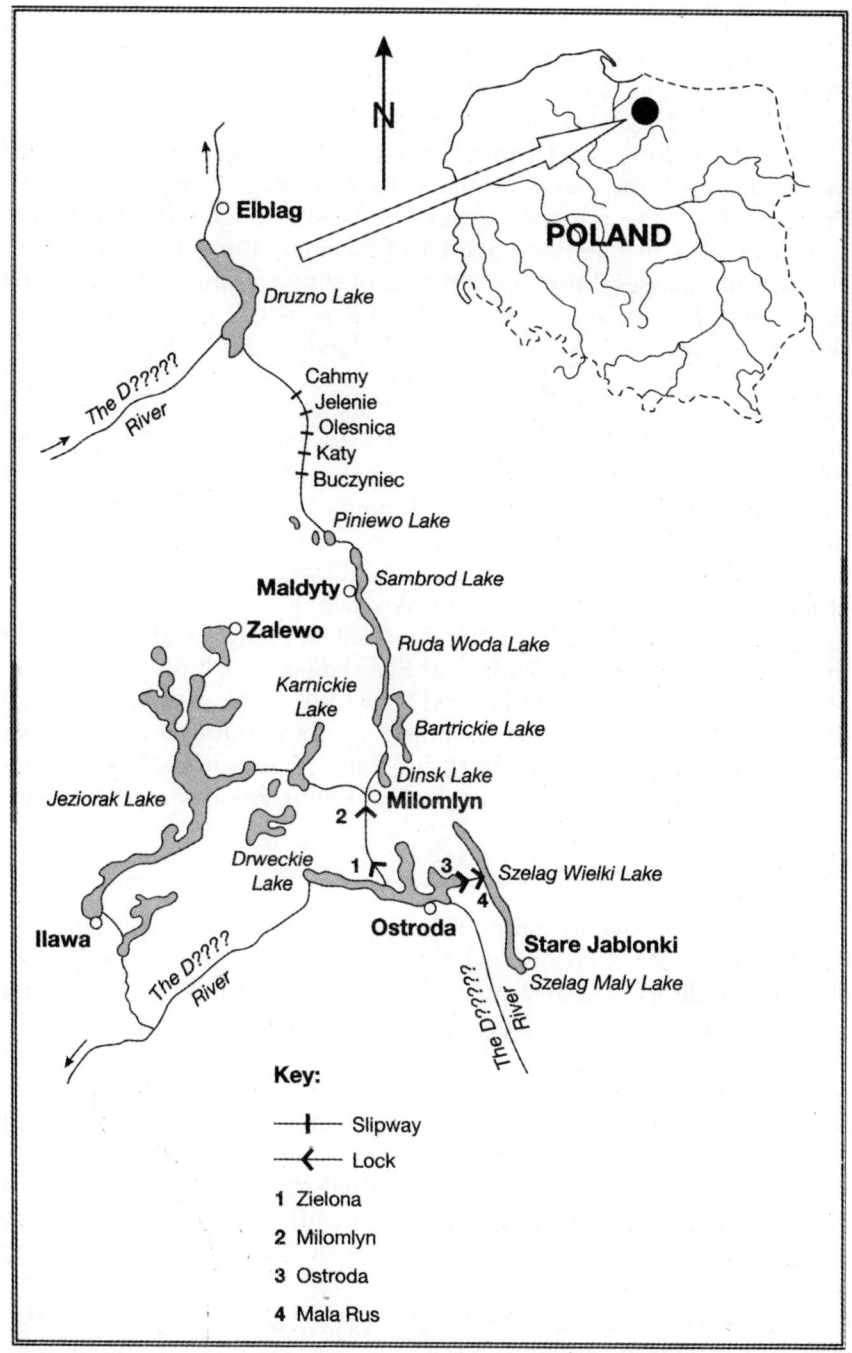

Figure 8.1 The Ostroda–Elblag Canal

The canal consists of three main sections 90 miles (144.5 km) long and several branches (Figure 8.1). The longest principal section, measuring 50 miles (80.4 km), joins Ostroda with Elblag (Baldowski 1998). Apart from being the longest stretch of canal, this section is also the most interesting part in terms of technical canal engineering and scenery. The section begins at the marina of the Ostroda–Elblag Navigation on Lake Drweckie, from where it goes to the north-west through the longest chain of ribbon lakes in Ilawskie Lake District (Ilinsk, Ruda Woda, Sambor, Piniewo) and passes through the system of slipways and locks described above and then to Druzno Lake. The main towns and tourist centres along this route are Ostroda, Pilawki, Milomlyn and Elblag.

The second branch of the canal leads from Ostroda via Milomlyn to Ilawa and is 29 miles (47.2 km) long. This section runs westward and passes through man-made ditches and three lakes – Karnickie, Dauby and Jeziorak. Straits and short streams connect Lake Jeziorak with other waterways. The west shore of Lake Jeziorak is covered by a vast forest complex hiding a large village, Siemiany, which is a popular destination among holidaymakers. The other holiday centres along this stretch of the canal are Ilawa, Karnity and Makowo.

The third section, which is 10½ miles (16.9 km) long, includes the route from Ostroda to Stare Jablonki and Staszkowo, and links lakes Szelag Maly, Szelag Duzy, Pauzenskie and Drweckie. The canal runs through diverse landscapes dominated by high forests and through two chamber locks at Mala Rus and Ostroda. Tourist and recreation centres are connected with the villages of Stare Jablonki, Katno, Zakatek and Mala Rus.

Tourist and Recreation Potential

The tourist and recreation potential of the Ostroda–Elblag Canal and the adjacent areas is created mainly by specific natural, landscape and climatic conditions. The area cut by the Canal is free from heavy industry, is mostly non-urbanised and is characterised by a large variety of landscape forms. There are natural or nearly natural habitats of plants and animals, including rare and endemic species. Landscape parks, nature reserves and other forms of protected land enhance the attractiveness of the region (Baldowski 2001a; Rakowski *et al.* 2002).

Different forms of landscape and their natural values help to create a microclimate which is typical of a given area (Januszewski 2001). In the temperate climate of the region, the summer season is relatively short and lasts from June to September. With other possible forms of tourist activity, such as angling and mushroom picking, included, the holiday season is usually regarded as starting on 15 May and continuing until 15 October. Except for some forms of fishing and hunting, the winter is considered a 'dead' season. However, this season could be promoted for tourism as there are good opportunities to develop winter sports, such as cross-country skiing, skating, sledging, ice-yachting, or rides in horse-drawn sleighs or dog-sledges.

Natural Environment

The area around the canal is one of the most interesting regions of Warmia and Mazury in terms of the attractions of the natural environment and landscape. It is a typical post-glacial landscape of varied relief, with a large diversity of morphological forms, lakes, springs, watercourses and other water bodies, and is rich in deep pine forests (Dabrowski & Szymkiewicz 2001). Lakes are one of the principal natural assets of the region. Table 8.1 identifies some of the characteristics of the lakes that belong to the water system of the Ostroda–Elblag Canal. The most important for water tourism are Lakes Drweckie, Jeziorak, Szelag Wielki and Szelag Maly.

Forests are also a significant part of the landscape that is attractive to tourists. The largest forests in the region are the Ostroda Forest (Puszcza Ostrodzka), covering 64,247 acres (26,000 ha). (Januszewski 2001) and Ilawskie Forests (Lasy Ilawskie), covering about 61,776 acres (25,000 ha). (Sawicki 2003). The wooded land is dominated by areas of coniferous and mixed forest with pine as the main species.

The importance of the region's natural assets is confirmed by the presence of many legally protected areas, including a number of significant nature reserves:

- Taborskia pines: the reserve protects a 230-year-old forest with the Taborska pine as the dominant species;
- mud turtle refuge;
- Black Lake – a dystrophic lake with a stand of *Isoëtes lacustris*.

The region of the Ostroda–Elblag Canal – the terrains adjacent to the canal and its nearest environs – is an area of protected landscape (Klim 1996). There are also three landscape parks: the Dylewskie Hills, the Ilawskie Lake District and the Elblag Elevations (Park Wzgorz Dylewskiech, Pojezierza Ilawskiego and Wysoczyzny Elblaskiej) (Rakowski et al. 2002), as well as many more nature reserves (forest, faunal and floral reserves).

The south-eastern part of the Ostroda commune contains the Landscape Park of the Dylewskie Hills, a unique upland area in this part of Poland, reaching 341 yd (312 m) above sea level. In winter, even mild ones, the Dylewskie Hills are ideal for winter sports. Another attractive feature of the hills is the vegetation, which consists of montane and submontane species of coniferous trees (Dabrowski & Szymkiewicz 2001).

The Landscape Park of the Ilawskie Lake District comprises three nature reserves: Lake Francuskie (a floral reserve to preserve a relic stand

Table 8.1 Description of the lakes located in the main sections of the Ostroda–Elblag Canal

Lake	Area (ha)	Maximum length (km)	Maximum width (km)	Maximum depth (m)
Ostroda–Elblag stretch				
Drweckie	870.0	15.50	1.10	22.3
Ilinsk	234.4	5.75	1.00	27.4
Ruda Woda	654.1	12.05	1.75	27.8
Sambrod	128.4	3.70	0.35	4.3
Piniewo	45.1	1.40	0.47	8:0
Druzno	1446.0	10.00	2.20	3.0
Ostroda–Ilawa stretch				
Karnickie	123.0	2.50	0.80	3.3
Dauby	62.5	1.90	0.49	3.7
Jeriorak	3219.4	27.45	2.35	12.0
Ostroda–Stare Jablonki stretch				
Pauzenskie	211.8	3.77	1.15	2.0
Szelag Maly	83.8	2.70	0.58	15.2
Szelag Wielki	599.0	12.50	0.90	35.5
Total area of lakes in the Ostroda–Elblag Canal route: 7677.8 ha				

Source: Janczak (1997); Januszewski (2001)

of the whortleberry willow), Dylewo (a forest reserve to protect a 115-year-old stand of Pomeranian beech) and part of the River Drweca Park (an aquatic park, established to protect the water habitat and to preserve many fish species inhabiting the Drweca, its tributaries and the lakes it flows through). The Landscape Park was established to preserve the natural, cultural and recreational assets of the lake district. The popularity of this region is due to its large number of lakes and woods, varying land relief forms and interesting plant and animal life. Rich in lakes and connected with the system of the Ostroda–Elblag Canal, the Ilawskie Lake District is dominated by water-sports tourism (sailing, canoeing, windsurfing, swimming, angling). The park is also well prepared to handle tourist traffic, with a network of tourist trails, marinas, water-equipment rentals and accommodation facilities. The park comprises three nature reserves: two bird sanctuaries at Gaudy Lake and Czernica and one water reserve at Jasne, which protects Lakes Jasne and Luba together, with woods and peat-bogs around the lakes (Dabrowski & Szymkiewicz 2001).

The Landscape Park comprises the northern parts of the mesoregion of the Elblag elevations and the south-east coast of the Vistula estuary. This is an area of exceptionally valuable natural and landscape assets. The park and its environs are an attractive place to pursue different forms of tourism. The Vistula estuary is suitable for sailing in summer and ice-yachting when it freezes in winter. Three nature reserves have been established in and near the park: Kadynski forest and beech woods of the Elblag Elevations preserve old beech woods, while Elblag Bay is a bird sanctuary which protects nesting and resting places of many species of water and marsh birds (Dabrowski & Szymkiewicz 2001).

The route of the canal crosses another ornithological reserve called Druzno Lake (covering an area of 7,466 acres (3021.6 ha)). This shallow water reservoir (an average depth of 2½ ft (0.8 m)) is heavily overgrown with plants. It is an important nesting and resting area for many species of water and marsh birds (Dabrowski & Szymkiewicz 2001). Druzno Lake has been chosen to join the EU nature conservation programme, Natura 2000, as a bird reserve and habitat site (Ministry of Environment, 2004). Other habitat sites are the Drweca river valley and Lake Karas, while the Vistula estuary and Ilawskie forests are protected areas for birds.

Cultural and Historic Factors

The anthropogenic assets of the region are less important than the to its natural environment. The main attraction is the canal itself and its facilities. This monument of technology is unique on the European, if not worldwide scale owing to its system of inclined slipways, the only one still in operation. Other objects worth interest include the town of Ostroda with its reconstructed 14th-century Teutonic castle and other historic buildings (Baldowski 2001b). Also lying on the canal route is the village of Milomlyn, which was established in the early 14th-century and has preserved some relics of its past, such as a 14th-century belfry, a corner tower and some houses from the 18th and early 19th centuries. Several attractive tourist destinations are also located near to the canal – e.g. Frombork, Malbork, the battlefield at Grunwald and monuments and other objects of interest in Elblag, Olsztyn and Gdansk.

Accessibility of Tourist and Recreational Areas

Accessibility defines the extent to which it is possible to utilise the existing tourist and recreational attractions. The canal can be reached quite easily. Its main section runs parallel to the E-77 road which connects

Warsaw with Gdansk. The canal's branch to Stare Jablonki and Ilawa is also well-connected with the road system. Stare Jablonki lies on the Olsztyn–Ostroda road and Ilawa is situated near the Ostroda–Grudziadz route. Stare Jablonki, Ostroda, Ilawa, Maldyty and Elblag have convenient railway connections with all major cities of the region: Olsztyn, Gdansk and Torun, as well as with Poland's capital city, Warsaw. The region also has good telecommunications access.

Tourist and Recreational Infrastructure

The main components of the tourist infrastructure of the region are holiday resorts, hotels and guesthouses, and camping and caravan sites. According to the 2002 statistical data, the total number of accommodation places (equivalent to beds) – (seasonal and all year) in the *voivodship* of Warmia and Mazury was 34,255. The number of tourists accommodated was 711,507 (Table 8.2). In the three administrative divisions (*powiats*) which the canal runs through (Elblag, Ilawa and Ostroda), 5105 beds were registered, of which 61% (3115 places) were in the *powiat* of Ostroda. Those places were used by 78,789 guests, including 9502 foreign tourists. In the three *powiats* mentioned, the total number of accommodation visitor nights reached 223,845. The highest number of visitors was recorded in the *powiat* of Ostroda with 49,106 overnight guests. Interestingly, the

Table 8.2 Accommodation facilities of the *powiats* (intermediate administrative divisions) of Elblag, Ilawa, Ostroda and the *voivodship* (province) of Warmia and Mazury in 2002

	Number of accommodation facilities	Number of beds	Tourists accommodated		Accommodation provided (nights)	
Powiats	Total, as of 31.07.2002		Total	Foreign tourists	Total	Foreign tourists
Elblag	11	484	12,761	1,386	20,226	2,926
Ilawa	25	1,506	16,922	1,932	61,378	4,164
Ostroda	44	3,115	49,106	6,184	142,241	20,459
Total	80	5,105	78,789	9,502	223,845	27,549
Voivodship of Warmia and Mazury	368	34,255	711,507	205,212	1,868,491	480,120

Source: Statistical Office of Olsztyn (2003)

lake region appears to attract proportionately fewer numbers of foreign tourists than the province as a whole.

An inventory of accommodation and tourist infrastructure undertaken for the Ostroda *powiat* by the Department for the Local Development and Promotion (DLDP) in 2004 identified 4585 accommodation places (beds), of which 1887 were available year-round (Table 8.3). More than half the beds available were seasonal, located mainly in resorts. However, the DLDP (2004) reported that fewer than half the places available in resorts and camp sites met the basic standards laid out for tourist accommodation (1047 out of 2315). The total number of places complying with the standard recommendations in tourism was 2501 out of 4585.

Water-based tourism also has specific infrastructure requirements. Water sports and activities require specific technical and navigational facilities – for example, landing bays and moorings, water-equipment rentals, workshops for boat repairs as well as points to refill fuel and water tanks, charge batteries and dispose of waste from yachts and boats. These facilities are mainly concentrated in the holiday centres on the canal route. Given the appeal of the natural environment in the canal region, it is also essential to respect the restrictions that regulate the

Table 8.3 Accommodation facilities available in the *powiat* of Ostroda in 2004

		Number of accommodation places (beds)		
Accommodation type	Number of providers	Available all year	Seasonal holiday resorts	Total
Hotels	16	1439	0	1439
Motels, inns	3	37	0	37
Guesthouses	7	139	23	162
Resorts, camp sites	18*	0	2315	2315
Agritourism farms	41	133	207	340
Youth and tourist hostels	2*	0	110	110
Rooms and apartments	18	139	43	182
Total	105	1887	2698	4585

*seasonal

Source: Department for the Local Development and Promotion at the Local Government Council in Ostroda (2004)

planning and construction of water tourism facilities and their utilisation. Given Poland's accession to the European Union, these restrictions should be harmonized and compatible with European legal regulations.

The negotiations with the European Union concerning the adjustment of the existing and future law on environmental protection to the EU legal system officially commenced in December 1999 and concluded on 25 November 2002 (Ministry of Environment 2003). The regulations on the protection of the natural environment bound by the EU were transferred to the Polish legislation by several acts and ordinances, including the Law on the Protection of Natural Environment (Journal of Laws 62/01, item 629) and the Water Act (Journal of Laws 238/02, item 2022). Local governments are obliged to execute these legal regulations. Due to the high investment costs, it is not always possible to comply with all the requirements imposed by EU laws by the deadlines set in the relevant EU directives – e.g. Directive 91/271/EEC on municipal sewage and waste-water treatment. As a result, Poland has applied for permission to have transition periods for the gradual implementation of EU policies and regulations. Because the financing of environmental impact statements for investment projects, even those executed by local governments, should not be based exclusively on state funds, it is possible that such projects will be financed by future users (Ministry of Environment 2003). Now that Poland is a member of the European Union, its government can apply for the co-financing of investment from the Structural Funds and Cohesion Funds.

A representative of the Directorate General for the European Commission of Regional Development visited the *voivodship* of Warmia and Mazury on 23–24 June 2004. The aim of the visit was to exchange and share the experience and know-how of the management and implementation of the Structural Funds and Cohesion Funds. The visitors saw some tourist attractions (e.g. the Ostroda–Elblag Canal) and some other projects that are likely to be co-financed from these funds, as well as the projects that have been implemented – for example, the waste-water treatment plant in Olsztyn (K. Swiatek, 2004, personal communication).

Water-based tourism

Water-based tourism in the Ostroda–Elblag Canal's region is primarily based on scheduled passenger cruises (also referred to as organised passenger cruises (OPS)) and cruises by private yachts and boats. Passenger cruises (PS) are provided by six passenger ships, of which five have a capacity of 65 persons, and a charter boat which can take 48 passengers. Private navigable vessels (PNV) consist primarily of sailing boats and yachts equipped with motor engines. Canoeing trips are not very common.

In order to identify the dynamics of tourism on water routes of the canal the number of operations executed at the locks and slipways was chosen as the most informative and reliable source of data over the period 1996–2002. The canal's routes were divided into two sections: the Region of the Slipways (the Buczyniec–Elblag section) and the System of the Warmia Lakes (the sections of Ostroda–Stare Jablonki, Ostroda–Piniewskie Lake, Ostroda–Milomlyn–Ilawa). Analysis of tourist movement in the Buczyniec–Elblag section was based on the number of total passages made at all the five slipways and in the System of the Warmia Lakes was performed on the basis of the total annual number of vessels using the four locks. The primary data were obtained from the Regional Water Management Office in Gdansk, from the registers of the hydrotechnical facilities held for each lock and slipway.

The number of times that the slipways were put to use to serve organised cruises increased from 2944 in 1996 to 4728 in 2000 (see Table 8.4). In the following two years this number slightly declined. The highest intensity in the movement of private boats and yachts was observed in 1999, with a total of 5957 operations of the slipways. Each time a slipway is put to use, one ship can be transported up or down the ramp. The smaller size of private navigable vessels mean that two or three boats or yachts can be carried each way on each occasion. According to Zwolinski (1992), a sailing boat has an average capacity of five persons, so it can be concluded that although the number of private vessels passing through the slipways was higher, the organised cruises on ships are

Table 8.4 Tourist flows on the Ostroda–Elblag Canal – the region of the slipways and the system of the Warmia Lakes in 1996–2002.

Years	Number of slipway operations		Number of lock operations	
	OPS	PNV	OPS	PNV
1996	2944	4958	653	1296
1997	3211	3820	656	1978
1998	3686	4308	545	1471
1999	4342	5957	667	1717
2000	4728	5563	720	1637
2001	4630	4884	617	2015
2002	4203	4620	602	2400

OPS: organised passenger cruises
PNV: private navigable vessels

more important for the total number of tourists using the facilities of the canal. In the seven years analysed, the number of sluiced large ships varied from 545 in 1998 to 720 in 2000, but did not show any distinct tendencies. However, the number of operations for private sailing boats and yachts increased nearly 100% – from 1296 in 1996 to 2400 in 2002.

Differences in the flow of tourism traffic on the water routes in the canal region in 1996–2002 were determined using the average annual number of passages through the locks and slipways (Table 8.5). The highest intensity of tourism flow in those years was recorded at the stretch between Buczyniec and Elblag, where it reached more than 1700 passages per slipway annually. This reflects the high attractiveness of the area associated with the localisation of the unique hydro-technical facilities and landscape attractions. Water traffic was much less intense in the region of the Warmia lakes and the highest number of vessels passed through the locks at Zielona and Milomlyn, which lie on the route to Ilawa and to the Region of the Slipways (up to 782 sluicing operations).

The seasonal character of water-based tourism in the study area is closely connected to the five-month-a-year cycle (1 May to 30 September)

Table 8.5 Annual tourist flows in the region of the locks and slipways in 1996–2002

Year	Number of passages								
	Slipway					Lock			
	Caluny	*Jelenie*	*Olesnica*	*Katy*	*Buczyniec*	*Milomlyn*	*Zielona*	*Ostroda*	*Mala Rus*
1996	1595	1577	1577	1577	1576	579	563	469	338
1997	1393	1396	1431	1458	1353	861	924	469	380
1998	1523	1552	1562	1613	1744	641	629	431	315
1999	1975	1944	1996	2112	2272	762	802	475	345
2000	1979	1997	2038	2,111	2166	775	770	477	335
2001	1785	1890	1891	1948	2000	777	793	673	389
2002	1749	1864	1557	1785	1868	1057	996	564	385
Mean (±SD)	1714 (±233)	1746 (±232)	1722 (±245)	1801 (±263)	1854 (±325)	779 (±154)	782 (±151)	508 (±83)	355 (±29)

of operation of the hydrotechnical facilities of the canal. The intensity of tourist flow shows some variation, both in terms of months and types of water activities (Table 8.6). Analysis was based on the number of times the canal's slipways were set in motion per month during 1996–2002. The highest seasonal variation is observed in the recreational activity of tourists travelling aboard sailing and motor yachts. As would be expected, the distribution of the OPS vessel traffic intensity over the months was more uniform. The peak of the tourist season for private yachts and boats occurs in July and accounts for 48% of the average annual traffic. The concentration of individual water-based tourism in July and August is a result of the fact that this is the Polish summer holiday season and optimum weather conditions usually prevail in these months.

The companies serving organised tourist traffic annually transport over 40,000 tourists. Their activity, however, clearly shows the seasonal character of the group tourism on which they focus. In May and June, these are organised groups of schoolchildren, while in September groups of

Table 8.6 Average seasonal distribution of tourist traffic in the region of the slipways of the Ostroda–Elblag Canal in 1996–2002

Month	Slipway operations			
	OPS*		PNV*	
	Mean monthly number	Percentage per month (%)	Mean monthly number	Percentage per month (%)
May	177	18.3 (±1.2)	28	2.7 (±0.2)
June	224	23.2 (±0.3)	150	14.4 (±1.8)
July	198	20.5 (±4.3)	496	47.6 (±0.4)
August	219	22.7 (±2.0)	317	30.5 (±0.4)
September	148	15.3 (±1.0)	50	4.8 (±0.9)
Total	966	100.0	1041	100.0

OPS: organised passenger cruises
PNV: private navigable vessels

foreign tourists, mainly from Germany, are predominent. It should also be noted that as a large proportion of tourist flow, especially on the Warmia lakes, omits the locks and slipways, its volume is difficult to assess. However, a potentially significant problem for future development of organised group tourism is that a cruise along the whole Ostroda–Elblag section of the canal takes about 11 hours and is too long, as well as possibly too monotonous, for many visitors. Therefore, shorter cruises should be developed and promoted. Such cruises may also be better integrated with land-based activities, given that the canal's water routes and facilities also enhance the region's landscape values, which are of importance for hiking, cycling, car driving, horse riding and other forms of recreation and relaxation.

Development of Tourism and Promotion of the Ostroda–Elblag Canal

Tourism is particularly important for the social and economic development of Warmia and Mazury, the region through which the canal runs. 'The Strategy of the Social and Economic Development of the Warminsko–Mazurkie *Voivodship*' (province) prepared for 2000–15 assumed that tourism would be a driving force of the regional economy (Board of the Warminsko–Mazurskie *Voivodship* 2000). Tourism is important for the regional economy partly because of the difficult situation in the local labour market. At the end of June 2002, the number of unemployed reached 27.9% of the labour market and in some *powiats* (intermediate administrative divisions) – e.g. Ostrodzki and Elblaski which lie in the area of the canal) exceeded 30% (Central Europe Trust Poland 2003). At the end of 2001, the number of registered unemployed in professions related to the hotel industry and tourism in the Warminsko–Mazurskie *voivodship* reached 676. It has been stated that every job place in the tourism sector creates 2.4 jobs outside the sector, with forecasts estimating that employment in the tourism economy should increase by 8800 people by 2007 (Central Europe Trust Poland 2003).

With its highly valuable natural and landscape attractions, and tourism and recreational activities, the canal region is a good choice for the development of tourism, especially water-based tourism. The canal, being a significant European example of working industrial heritage, is an important component of this potential (Board of the Warminsko–Mazurskie *Voivodship* 2000). Among the actions undertaken to take advantage of the region's tourist industry potential are: maintaining the technological operation of the canal and its use as a waterway connecting Jeziorak lake and the adjacent areas, increasing tourist activity in the Vistula estuary and Druzno lake and developing and managing tourism in the Drweca river catchment. In addition, the international E-77 road (a European road designation), which runs along the canal, as well as the inland navigational waterway consisting of the System of the Warmia Lakes and canal, are important aspects of the infrastructure of the region.

The development of tourism in the whole *voivodship* as well as in the area of the canal, lies in the interest of local and regional government authorities. In 1997, an Association of Communes of the Ostroda–Elblag Canal and Ilawskie Lake District was established (Journal of Laws of the Olsztyn *Voivodship* 1997). At the time of writing, the association had seven members. The tasks of the association are to promote the canal, the Ilawskie Lake District and the member communes, and to create conditions for the management of the canal and the adjacent areas for the development of tourism. The association has undertaken a number of actions. Since 2001, attempts have been made to nominate the canal a UNESCO World Heritage site. In 2002 the association commissioned a film about the canal to be used for promotional events. An Internet website is being constructed, on which both the canal and the member communes will be presented. However, the most significant initiative of the association resulted in the creation of 'The programme for the development of tourism in the area of the Elblag Canal and the Ilawskie Lake District' (S. Panczuk, 2004, personal communication). In August 2002, an agreement was signed between the local government of the Warminsko–Mazurskie *voivodship*, the Association of the Communes of the Ostroda–Elblag Canal and the Ilawskie Lake District regarding implementation of the programme (Elblag Canal and tourism 2004). Eighteen communes from the three *powiats* – Elblag, Ostroda and Ilawa – which have economic linkages to the canal, joined the initiative. A total of 117 projects were submitted in May 2004 to the Association of the Communes of the Ostroda–Elblag Canal and the Ilawskie Lake District as the owner of the programme (S. Panczuk, 2004, personal communication; Elblag Canal and Tourism 2004). The total value of the tasks according to the projects is about €6.8 million. The tasks planned for implementation in 2004–06 amount to €4.7 million. The main objective of the programme is to activate the region economically through the development of tourism. The priorities of the programme are:

(1) Revitalisation of the canal, comprising ten projects, five of which aim to repair the canal.
(2) Development of tourism (54 projects).
(3) The technological infrastructure – roads and environment (53 projects).

The projects require further development and specific knowledge of the legal acts on protection environment both in Poland and in the European Union.

Another task related to tourism (Board of the Warminsko–Mazurskie *Voivodship* 2000) is to create regional and local tourist organisations (RTO and LTO), in order to adjust the organisation of tourism in Poland to the situation that exists in other European countries. An LTO is being created in Ostroda. This is a non-government organisation to which local governments of the communes connected with the canal and main tourist operators in the area have acceded. The main statutory aim of the LTO concerns executing marketing and promotional activities for the region by participating in tourist trade fairs, promotional events or specialist trainings; by publishing catalogues, brochures and other promotional materials, and by taking part in the elaboration of tourist development and tourist infrastructure modernisation plans.

Publicity and advertising campaigns promoting tourist destinations are essential for tourism development. The canal is promoted at Polish and foreign tourist trade fairs. Brochures and information packs are published in several languages. Meetings with journalists and photographic exhibitions are also organised on a regular basis – e.g. an exhibition of photographs by Piotr Placzkowski called 'The Ostroda–Elblag Canal' shown in Ostroda and Milomlyn in June and July 2004). Information about the canal (also in foreign languages) can be found on websites of the local communes and *powiats*, and on the website of the Ostroda–Elblag Navigation – e.g. www.zegluga.com.pl). In September 2004, the Ostroda–Elblag Canal Navigation was among the winners of the competition organised by the Polish Tourism Organisation (PTO) and was granted a PTO certificate for the Best Tourist Product 2004 (Polish Tourism Organisation 2004). This award involves intensive promotion of the product both in Poland and abroad. Unfortunately, the E-77 road (Warsaw–Gdansk) that run along the main branch of the canal and connects its main points, has very few signposts, indicating the proximity to the canal. Other roads running along the canal or crossing the canal's branches are poorly signposted, too. Few signs showing directions to the canal can be found in the towns and villages near the canal. Undoubtedly, the issue of signposting will be a major issue in the future.

The Ostroda–Elblag Canal serves as a significant component of the economy of the Warminsko–Mazurskie *voivdoship* and, given the level of unemployment in the region, is likely to assume greater importance in terms of tourism-related economic development in the future. However, the high costs of constructing and maintaining a tourist and recreational infrastructure, as well as the requirements for preserving the natural environment are forcing investors to look for ways of extending the short summer tourist season in the canal's region. Nevertheless, to adapt the tourist accommodation and infrastructure to the autumn and winter season, primarily by providing heating, will require significant investment, as will the need to improve the quality of accommodation as well as promotion and advertising. Although future tourism developments will undoubtedly have economic benefits, such tourism and recreational developments will need to be undertaken in a manner that ensures the conservation of the region's natural environment and landscapes on which its overall attractiveness is based.

References

Bakowska, M. (1996) Engineering marvel offers rare views. *The Warsaw Voice* 29, 404.
Baldowski, J. (1998) *Mini Tour Guide, The Ostroda–Elblag Canal*. Gdansk: TESSA.
Baldowski, J. (2001a) The region's tourist attractiveness. The geographic environment. In *Warmia and Mazury – An Illustrated Guidebook* (pp. 7–10, 19–24). Olsztyn: The Mazury Photographic and Publishing Agency.
Baldowski, J. (2001b) Description of some towns and villages. In *Warmia and Mazury – an Illustrated Guidebook* (pp. 63–294). Olsztyn: The Mazury Photographic and Publishing Agency.
Board of the Warminsko–Mazurskie *Voivodship* (2000) Strategy of the social and economic development of the Warminsko–Mazurskie voivodship. Available at www.amicro.pl/wm/Str/Str.htm.
Central Europe Trust Poland (2003) Labour and Tourist Market Analysis. Human Resources Development – Training for the Employees of newly established regional Tourist Information Centres, Warminsko – Mazurskie Voivodship, Poland. Project financed by EU PHARE Funds and Polish Budget Funds, Warsaw.
Dabrowski, S. and Szymkiewicz, M. (2001) Natural environment. In *Warmia and Mazury – An Illustrated Guidebook* (pp. 25–62). Olsztyn: The Mazury Photographic and Publishing Agency (in Polish).
Department for the Local Development and Promotion at the Local Government Council in Ostroda (2004) Programme of the Local Development for the *Powiat* of Ostroda. Annex to the Act number XXII/139/2004 of the Ostroda *Powiat* Council, of 19 July 2004 (in Polish).
Elblag Canal and Tourism (2004). Caring for the economic development (in Polish). *Głos Elblaga*, 97, 1.
Janczak, J. (ed.) (1997) *The Atlas of Poland's Lakes vol. II. Lakes of the Catchments of Pomerania Rivers and the Lower Vistula Catchment* (in Polish). Bogucki. Wydawnictwo Naukowe: Poznan.
Januszewski, S. (ed.) (2001) *Historic Monuments of Industry and Technology in Poland. 5. The Ostroda–Elblag Canal* (in Polish). Wroclaw: The Foundation of the Open Museum of Technology, Art and Advertising Studio TAK.
Journal of Laws of the Olsztyn Voivodship (in Polish) (1997) 31, item 440.
Klim, R. (1996) Historic and natural values of the Ostroda–Elblag Canal (in Polish). *The Amber Trails* 3, 10–12.
Kondracki, J. and Ostrowski, J. (1973–1978) *The National Atlas of Poland* (in Polish). Wroclaw – Warsaw – Cracow – Gdansk.
Ministry of the Environment (2003) The EU requirements on the protection of natural environment – practical information for local governments. Available at www.mos.gov.pl/integracja_europejska/wymagania.pdf.
Ministry of the Environment (2004) *Natura 2000, the European Ecological Network. Lists of Areas Submitted by Poland to the European Commission*. Available at www.mos.gov.pl/1strony_tematyczne/natura2000/lista_obszarow/index.shtml.

Polish Tourism Organisation (2004) A PTO certificate for the best tourist product. Available at www.pot.gov.pl/certyficat.asp.

Rakowski, G. (ed.) Smogarzewska, M., Janczewska, A., Wojcik, J., Walczak, M. and Pisarski, Z. (2002) *The Landscape Parks in Poland* (in Polish). Warsaw: Institute of Environmental Protection.

Sawicki, G. (2003) *Bird Sanctuary – the Ilawa Forest* (Polish – English version). Warsaw: Wydawnictwo Naukowe AKSON.

Statistical Office of Olsztyn (2003) *Voivdoship of Warmia and Mazury* (Polish – English version). Statistical Yearbook. Olsztyn.

Zwolinski, A. (1992) *Evaluation of the Tourist and Recreational Assets of Artificial Water Reservoirs* (in Polish). Warsaw: Dzial Wydawnictw Instytutu Turystyki.

Chapter 9
Finnish Boaters and Their Outdoor Activity Choices

TUIJA SIEVÄNEN, MARJO NEUVONEN AND EIJA POUTA

Introduction

Finland is rich with coastal areas, lakes and waterways that can be used for recreational purposes. Boating is a popular outdoor activity among the Finnish people, almost half of whom participate in some kind of boating activity and has access to a boat on a regular basis (Pouta & Sievänen 2001). A study of the participation of Finnish people in rowing, motorboating with a small or large boat, canoeing and kayaking, sailing with a small or large boat, windsurfing and jet-boating was carried out in the National Recreation Demand and Supply Inventory (LVVI) for the first time in 1998–2000 (Sievänen 2001). Statistics resulting from the study give some indication that the Finnish people's participation in leisure activities on the water has increased since the early 1980s (Liikkanen et al. 1993). A similar trend has been observed in the United States, where more accurate measurements of outdoor recreational activities have been made since the 1960s. In 1994–5 24% of the American population participated in some kind of boating activities, but by 1999–2000 this had increased to 36%. Participation in some boating activities such as sailing has increased only slowly, but other activities such as canoeing and kayaking have substantially increased their popularity in recent decades (Cordell et al. 1999; Green et al. 2003; Cordell et al. 2004). In Canada, participation rates are similar: 10% participate in canoeing, kayaking or sailing, and 9% in motorboating (DuWors et al. 1999). As an example of boating activities by Europeans in the Netherlands 14% of the population participates in boating activities such as sailing, rowing, canoeing and surfing (Statistics Netherlands 1997; Sievänen et al. 2000; see also Chapter 7, this volume).

As boating is one of the main outdoor activities in summertime in Finland there are many services for boating are provided by municipalities, sailing clubs and private entrepreneurs. There are about 600 harbours in Finland that are managed for recreational boating and which provide a wide array of services. There are also a number of less well-equipped harbours in lake districts, coastal areas and the archipelago. Access and temporary landing opportunities follow the country's traditional concept of 'everyman's rights.' However, in providing and developing public services for boaters and also commercial tourism, it is important to know the activity patterns of various types of boaters.

This chapter focuses on boaters and their participation in outdoor recreation in general. Some theoretical approaches are reviewed in order to offer a framework for interpretation of the empirical results. The aim of the chapter is to describe the recreational activity patterns of boaters in Finland and provides information on boaters' behaviour in terms of frequency of their participation and boating trips. The chapter also offers information on the importance of boating among finns, possible changes in participation and the need to invest in boating services by both public agencies and private enterprise.

Factors Affecting Choice of Activity: Recreation Motives and Socioeconomic Characteristics

Many factors contribute to individuals' varying perceptions of leisure, most which are related to an individual's social environment and personality (e.g. Bishop 1970; London et al. 1977). This psychological approach suggests that a person's decision to participate in an outdoor activity is based on the desire to satisfy a need. There are several ways to obtain such satisfaction. One activity can be substituted with another in order to get similar satisfaction. This suggests that individuals tend to choose activities that fulfil the same types of needs. The psychological basis of activity choice leads to an activity package of similar need satisfaction. Studies of activity choice have been carried out in recreation areas (McCool 1978) or have focused on a recreational trip of one or more days (Bristow et al. 1992). The approach of the research on which this chapter is based is more activity oriented in general – i.e. which outdoor activities do boaters seem to choose as part of their leisure lifestyle?

One of the classic studies of activity preferences was undertaken by Hendee et al. (1971) who developed a typology of outdoor recreational activity preferences based on an interpretation of the meaning of preferred activities. Preferred activities refer to the benefits or the source of satisfaction gained when people participate in a particular recreational activity. There are five categories of meanings into which recreational activities can be classified. The appreciative-symbolic category emphasises the nature values of the recreation environment. The extractive-symbolic category represents activities that offer playful extraction of commodities or trophies. The passive-free-play category includes activities that offer, most importantly, relaxation and rest. The sociable learning category focuses on satisfying communication needs and fostering a feeling of togetherness. The active-expressive class stresses the opportunity to use physical fitness and special skills when participating in recreational activities. In Hendee *et al.*'s (1971) study, quiet boating (rowing) and canoeing were mentioned as examples of passive-free-play and boat racing as an example of the active-expressive category. McCool's (1978) study, using this typology, classified sailing in the active-expressive category. Different types of boating activity offer various benefits and ways to satisfy different needs. This leads to the assumption that individuals, whose activity packages differ, participate in different type of boating activities.

Hendee *et al.*'s (1971) typology combines the motives and benefits identified by recreation research in later years surprisingly well (e.g. Driver *et al.* 1991). Boating motives are varied and dependent on the type of boating, the experience of the boater and boating opportunities in general (Bowes & Dawson 1999). Most of the studies of boating motives and benefits concern canoeing and other types of river recreation (Lime 1986). The most important motives are the opportunity to be in close contact with nature and to enjoy the scenery. Getting away from daily routine and being together with family and friends – common motives for all leisure activities – are also strong motives for boating. Tinsley and Johnson's (1984) study refers to canoeing as a way to express oneself, separating leisure from work and everyday life. It seems that boating offers two specific dimensions of experience that are not easily found in ordinary life – nor are they easily found in other outdoor activities. These are to experience peace and calm on the one hand and to experience excitement on the other (Lime 1986).

An example of the typologies of outdoor recreationists is presented by Cordell *et al.* (1997a). In their typology, there are seven different types of outdoor recreation participants. The 'sports hogs' is a group representing a very sporty lifestyle and the members of it have many types of boating activities in their activity packages. A group described as 'the bass club' favours motor-boating, together with fishing – mainly angling. Other groups are either the less nature-oriented 'fitness buffs', or are less keen on physically demanding outdoor activities such as 'nature lovers', 'fish and hunt avids', 'passives' and 'do nothings'. Members of the 'sporty' group are young and their education and income levels reflect the average levels of the whole population. The group of motor-boaters in 'bass club' represents older age groups and their level of income is below average.

Age, social status and education are most often found to be the best background factors to explain differences in people's choice of recreational activities. However, it is also obvious that socioeconomic factors alone cannot explain participation or non-participation in boating:

recreational behaviour patterns or recreational participation styles are always the result of a combination of factors. Indeed, boating activities that are physically demanding, such as canoeing, kayaking, rowing, and rafting, tend to have declining participation rates as people grow older. Motor-boating, which requires less exertion, is popular for people of all ages. Participation in boating also increases as household income increases, which is understandable because boats are often expensive. The same applies to education, as education and income level are often correlated. Boating participation also tends to increase among households of more than one or two members (Cordell et al. 1997b).

In order to get a more holistic picture of behaviour patterns and explanations of recreation styles, it is necessary to look for support from behavioural theories. There are two basic lines of behavioural science that can contribute for the purposes of this analysis. The theory of way of life in sociology provides a background from which the influence of social groups and social structure on individual recreation behaviour can be examined (e.g. Bourdieu 1984; Roos 1987). The way of life theory offers an interpretation using education, social status and income as variables in the empirical analysis. The way of life theory uses information about an individual's resources, life choices and values, which are all factors that may direct recreation behaviour. The same factors affect choice of recreational activities and recreation styles, such as frequency of participation, choice of recreation sites and social groups. In general leisure literature, the same theoretical approach is often discussed with regard to the concepts of lifestyle (e.g. Veal 1993). These approaches of explanations of leisure behaviour are used here to support the interpretation of boater's choices of outdoor activities.

Another line of theory, based most strongly on the factor of age, is the sociopsychological theory of life course or life cycle (e.g. Elder 1975; Rapoport & Rapoport 1975). Most individuals go through the same stages in life from childhood to old age and experience similar life events such as education, family life and career. Disposable free time and physical health, which vary greatly in the course of a person's life, are important resources for recreational participation (e.g. Kelly 1974; Liikkanen et al. 1993).

Combining these approaches to reach an interpretation of an individual's recreation behaviour provides a basis from which to look at the recreational activity packages of boaters. It can be assumed that the chosen recreational activities reflect the chosen leisure lifestyle, which includes both the choice of activities as well as the choice of recreational environments (Ditton et al. 1975). It is suggested that people choose the recreational environment (for example, water) first and then the recreation (e.g. boating) style follows, within the limits of overall lifestyle. Naturally, the supply of recreation (e.g. boating) opportunities is an influential factor in directing interest and behaviour.

Data and Methods

The results presented in this chapter are based on the population-based data collected in the context of the Finnish national outdoor recreation demand-and-supply inventory (LVVI) (see Sievänen 2001; Pouta & Sievänen 2001; Sievänen et al. 2003). The data was gathered in two phases through telephone interviews and a postal questionnaire in 1998–2000. These surveys were targeted at Finns in the age range of 15 to 74 (Virtanen et al. 2001). The total sample size was 12,649 persons. The Survey Research Unit of Statistics Finland collected data between August 1998 and May 2000. Interview data was gathered from 10,651 respondents (84% of those sampled). Information relating to participation in boating activities and number of days spent boating are based on the interview data. Information about the respondents' other outdoor activities was also obtained from this data. A follow-up to the LVVI study is not planned to occur until some time in the period 2008–20.

The postal questionnaire was sent to those respondents who were willing to answer it. Two-thirds (65% or 5535 persons) responded to the questionnaire. Information about boating trips is based on this data. The observations from the survey were used to form groups in correspondence analysis and to analyse the group characteristics (n = 2737). The boaters were divided into five groups, following the rareness of their activity and the frequency of participation (excluding occasional participants): sailing cruisers (n = 139) were selected first, then motor cruisers (n = 366); the next groups were canoers (n = 291) and small motor boaters (n = 888) and finally rowers (n = 1053). Boaters' participation in outdoor activities was analysed using multiple correspondence analysis (see Greenacre 1984). From about 30 outdoor activities, 20 differed statistically between the boater groups and were chosen in the correspondence analysis (chi^2-test, p-value <0.05). Differences in socioeconomic variables between boater groups were also compared using cross-tabulation and the chi^2-test. Analysis of variance was used for the factors describing boaters and boating behaviour. Multiple comparison tests were made using Tukey's test.

Boating Groups and Recreational Activity Choices

Participation in boating and boating trips

Almost half (47%) of the population had participated in boating in the course of one year. The most popular type of boating was rowing, in which almost 40% of the population participated, and 76% of boaters had used a rowing-boat to some extent. Small motor-boating was also a common activity, with 20% of the population and 44% of boaters participating. Canoeing was less popular (5% of the population and 11% of boaters). Boating with larger boats with the option of cruising was also popular with a minority of the population. Motor-cruising was the recreational activity of 7% of the population and 13% of boaters. Sail-cruising interested 3% of the population and 5% of boaters. Boating frequency varied between the different types of boating. The average frequency of participation per year was 17 times for rowing, 18 for small motor-boating, 7 for canoeing, 15 for motor-cruising and 11 for sail-cruising.

An average day boating trip lasted for six hours when started from a permanent home and three hours when started from a holiday home. The average distance to a boating site (harbor or similar) was 2½ miles (4 km) (median) from a permanent home. Day boating trips involved only minor costs – €20–30 per day trip (median). Overnight cruising trips lasted an average of six days, at a cost of about €73. For most boaters, the trips themselves are not so costly; the bulk of boating expenses are incurred in buying, maintaining and storing the boat. Harbours have relatively low fees and in many cases mooring is governed by the country's traditional concept of 'everyman's rights.'

Participation in outdoor activities by boater groups

The five boating groups – rowers, small motor-boaters, canoers, motor-cruisers and sailing-cruisers (yachts) – showed remarkable differences in their recreation participation patterns (Sievänen et al. 2003). Correspondence analysis suggests that the boater's activity profile differs between the boating groups. A two-dimensional solution is presented in Figure 9.1. The first (horizontal) dimension represents the diversity of a selection of outdoor activities chosen in each boating group. The more towards the left of the figure the boater group is, the smaller the number of recreation activities the boater tends to have; the more to the right the boating group is, the larger the number of recreation activities included in the boater's activity package. The second (vertical) dimension describes the type of activity selected, in terms of its image as a contemporary or as a traditional activity. Rowers are the furthest to the left regarding of diversity of activities, but are neutral in terms of their choice of traditional–contemporary style of activities. The rowers group participated most actively in wild berry picking compared to other boaters and they were the most passive group in terms of participating in recreational activities in general. Some 70% of rowers participated in wild berry picking (Table 9.1). The small motor-boaters as well as the motor-cruisers are in the middle of Figure 9.1. Small motor-boaters are the most active participants in fishing and hunting activities. Motor cruisers' activities can be considered both traditional and contemporary. Snow-mobiling, fishing with nets and backpacking are the most popular choices among motor-cruisers, compared to other boaters, but the group also tends to participate in a wide variety of activities. Canoeists are furthest to the right regarding diversity of activities. Canoeists are the group with the largest variety of recreational activities and the one, which is also modern in its choice of activities. Canoeists are the most active participants in many of the activities studied. Typical activities for that group are backpacking, camping, long-distance cycling, diving and all types of skiing, but also fishing, hunting and snow-mobiling. It may be that the canoeists group includes two separate recreation groups. The sailing-cruisers are positioned left in Figure 9.1, but highest in terms of selecting a traditional-contemporary style. Sailing-cruisers to have the most contemporary activity choices: they favour picnicking, downhill

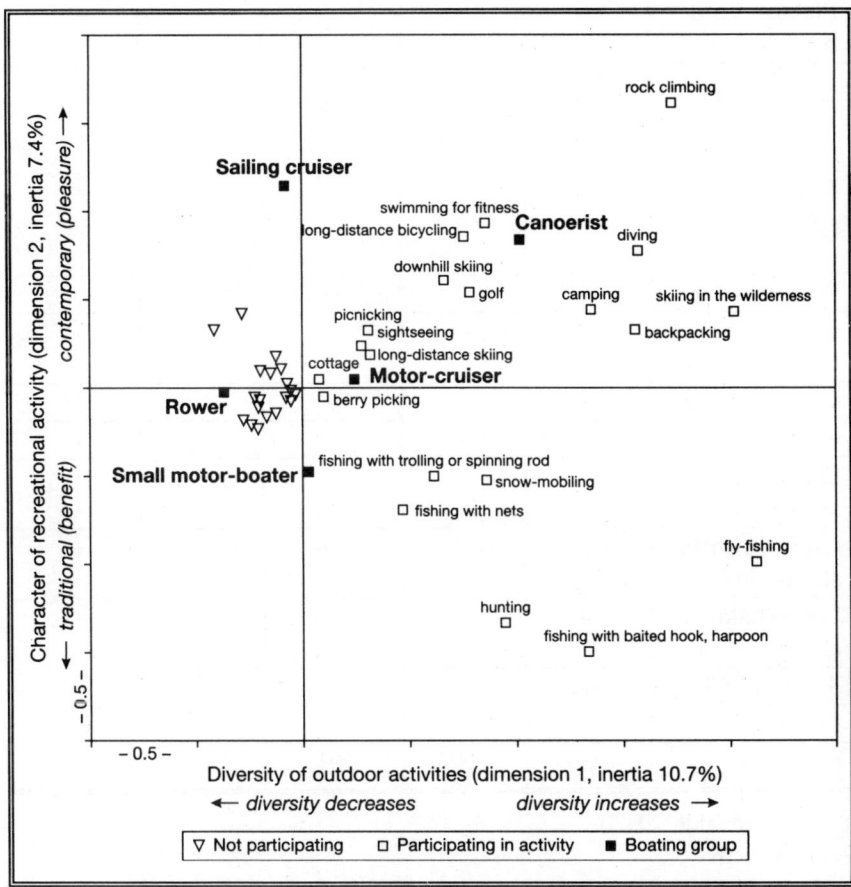

Figure 9.1 Boating groups, and groups of participants and non-participants in different recreational activities

Table 9.1 Boating groups described according to choices of other recreational activities

	Rowers	Small motor-boaters	Canoeists	Motor-cruisers	Sailing-cruisers (yachts)
Portion of boaters, %	38.5	32.4	10.6	13.4	5.1
Recreation activities participated in least and most					
Picking wild berries	++				—
Hunting	—	++	+		—
Fishing with nets		++	—	+	—
Fishing with spinning or trolling rod	—	++	+		—
Backpacking	—	-	++	+	
Long-distance bicycling	—	—	++		+
Camping	—		++		
Diving	—	-	++		+
Downhill skiing	—	-	++		+
Cross-country skiing			++	-	+
Long-distance skiing	—		++	+	—
Snowmobiling	—	+	+	++	—
Picnicking	—				++
n = 2737	1053	888	291	366	139

(chi^2-test, p-value <0.05)
++: most active
—: Least likely to participate in outdoor activity

skiing, diving and long-distance cycling. The only activity which separates them from other boating groups is picnicking, in which they participate most actively. With regard to most outdoor activities, the sailing-cruisers were the least active group.

The number of other outdoor activities that the boaters participated in differed between boating groups (Table 9.2). Rowers listed fewer activities and canoeists listed more activities than other boaters. Also, the amount of money used for outdoor recreation per year varied between groups. Rowers and canoeists used less money compared to others, particularly motor boaters. There were no differences between boating groups in terms of distance to the boating site or number of vacation days.

Boaters' socioeconomic characteristics are to some extent different between boating groups (Table 9.3). Rowers and small motor-boaters are older than others; they are often farmers or manual workers and their educational level is below average. They more commonly live in rural communities in northern and eastern Finland. Canoeists are distinctly young and students. Motor- and sailing-cruisers are both managerial and blue-collar employees and sailors tend to have a higher education and income than other boaters. Cruisers mainly live in towns in southern Finland.

Conclusions and Discussion

The study of Finnish boaters and their choices of outdoor recreational activities identifies two interesting points: there is a tendency that certain outdoor activities create 'outdoor recreation packages' and that those activity packages also tend to be related to certain of the groups population. Different boating groups participate in distinctly different kinds of outdoor activities and the typical socioeconomic characteristics of each group differ to some extent as well. Canoeists seem to be the most active group, participating in many kinds of outdoor activity, particularly contemporary, physically demanding ones, which can be interpreted as falling into the active-expressive category in Hendee et al.'s (1971) typology. Canoeists formed the youngest group and they were mostly students. Youth also seems to apply to American canoeists (Green et al. 2003). Canoeists had the lowest number of boating times per year, which is probably due to their high participation rate in many other activities. Sailing-cruisers are close to canoeists in their array of activities, but they participate in fewer activities. This may be explained partly because sailors are older, but also because sailing itself is a time-consuming activity. American sailors also belong to middle-aged groups, are middle- or high-income earners and mainly urban dwellers (Green et al. 2003). Finnish canoeists' and sailors' socioeconomic profiles are the most

Table 9.2 Number of boating days or times, distance to boating site, number of vacation days and amount of money used for outdoor recreation in the household on average in different boating groups.

Factors describing boaters and boating behavior	Rowers	Small motor-boater	Canoeist	Motor-cruiser	Sailing-cruiser	One-way analysis of variance
			mean (standard error)			F-test (p-value)
Number of boating times per year[1]	16.8 (0.73)	18.4 (0.81)	6.5 (0.53)	17.3 (1.23)	12.9 (1.60)	–
Distance to boating site from permanent residence (km)	6.0 (0.52)	6.4 (0.65)	5.5 (0.96)	5.1 (0.76)	5.5 (0.79)	0.49 (0.7409)
Number of recreational activities participated[2]	14.2[a] (0.17)	16.1[b] (0.20)	20.1[c] (0.45)	17.1[b] (0.41)	16.5[b] (0.58)	54.4 (<0.0001)
Number of vacation days per year	32.9 (1.51)	33.7 (1.69)	37.9 (2.27)	35.1 §(2.48)	37.2 (2.90)	0.87 (0.4797)
Amount of money used for outdoor recreation (€/year)[2]	167[a] (43.2)	477[c] (74.4)	198[a] (54.3)	613[c] (28.9)	434[a] (87.3)	6.05 (<0.0001)

[1] Number of boating times of boaters in the group to where they were classified.
[2] Differences of group means are described by symbols a, b and c (Tukey's test). Different letter symbol indicates that means differ at p<0.05 and the same letter that they do not differ at p<0.05.

Table 9.3 Description of boaters according to their socioeconomic characteristics

Socioeconomical variable	Rowers	Small motor-boaters	Canoeists	Motor-cruisers	Sailing-cruisers
Age class	45–64, 65+	45–64, 65+	15–24		24–44
Social status	Farmer, manual worker	Manual worker	Student	Lower level employee	Upper level employee
Education	Lower secondary general	Lower secondary general or upper secondary vocational	Upper secondary general	Upper secondary vocational, university	University
Income	Low			Middle	High
Size of community of residence	Rural village	Village		Town	City
Region of community of residence	Eastern, central, northern	Eastern, central, northern	Central	Southern, Uusimaa province	Uusimaa province
Access to equipment and facilities		Small boat hunting fishing	Camping hunting	Boat with beds	Boat with beds

Note: Differences between socioeconomic variables and boating groups are statistically significant (chi^2-test, p-value <0.05).

similar, compared to other boaters, but they are at different stages of their lives.

Typical activity choices of motor-cruisers are hiking, long-distance skiing and fishing, and also snow-mobiling. These activities can be classified into two categories in Hendee et al.'s (1971) typology. They are both active-expressive, but also sociable learning kinds of activities. Motor-cruising can be described as a sociable learning or a socialising activity, because motor-cruisers typically spend more time in harbours than sailors.

Small motor-boaters and rowers are most typically classified in Hendee's et al.'s extractive symbolic category. Both favour traditional outdoor activities such as fishing, hunting and picking berries. Rowers may also fall in the passive-free-play category since often they have few other activities and rowing itself can be done in a rather relaxing or less physically demanding way. Both small motor-boaters and rowers most typically live in rural areas in eastern or northern Finland, and belong to an older and less educated population group (Figures 9.2 and 9.3). Both of these types of boating are also typically associated with spending time at a summer cottage and thus people who are categorised into other boating groups in this study often also participate in these types of boating as well. That is a limiting factor to the interpretation of the results of this study. Also, Finnish motor-boaters are the most difficult group to compare to American boaters because of the close relationship between motor-boating and summer cottage use in Finland. In the United States, motor-boaters are mainly urban, middle-income, middle-aged people (Green et al. 2003).

Based on the outdoor activity choices of the boaters, it seems that the groups fall into one of two different lifestyle categories: Rowers' and small motor-boaters' activity choices reflect more traditional, rural values and lifestyles. Canoeists and sailing-cruisers, although at different stages of their lives, represent a more urban, contemporary lifestyle. This chapter offers a description of boaters and their outdoor activity choices, which are part of their way of life as a whole. However, it does not provide an explanation of boaters' profiles. It is a challenge in terms of data and methods to find all the social, psychological and economic factors that would contribute to a deeper understanding of the boating behaviours of the Finnish population. Nevertheless, the description of boaters' activity choices and the basic socioeconomic characteristics of the different groups presented in this chapter offer a basis for developing recreational services for boaters and also for water-based tourism and leisure boating in Finland.

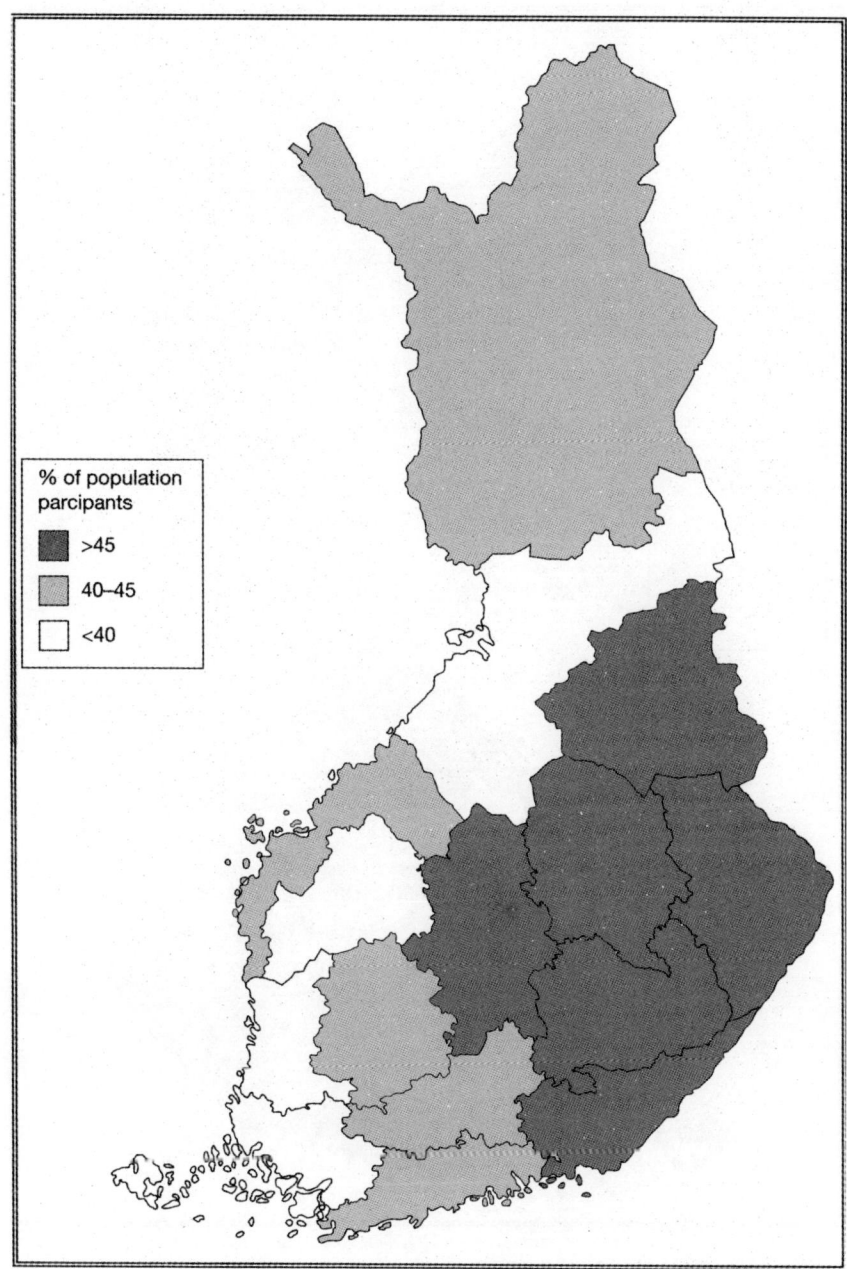

Figure 9.2 Percentage of the Finnish population participating in rowing or small-craft motor-boating as a recreational activity by region

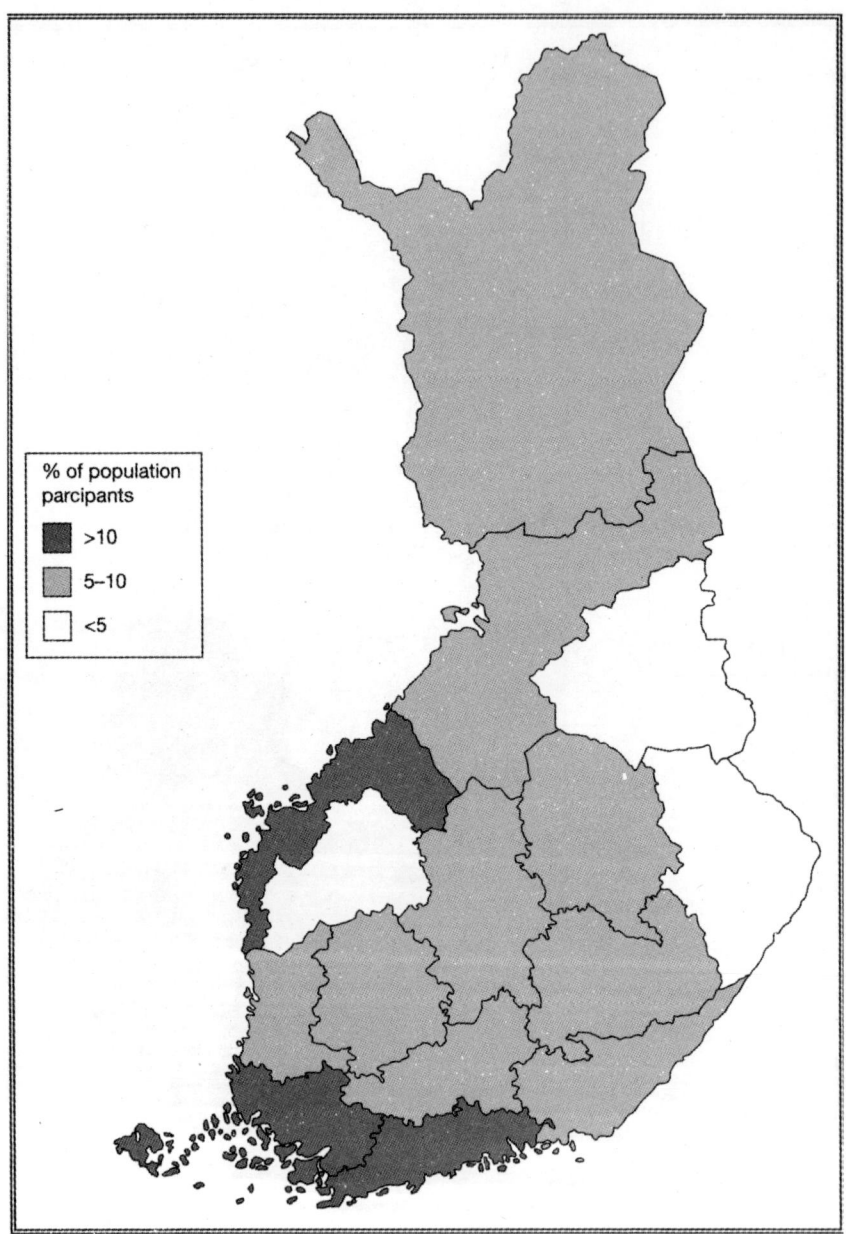

Figure 9.3 Percentage of the Finnish population participating in sailing or motor-cruising as a recreational activity by region

References

Bishop, D.W. (1970) Stability of the factor structure of leisure behaviour. Analyses of four communities. *Journal of Leisure Research* 2 (3), 160–70.
Bourdieu, P. (1984) *Distinction. A Social Critique of the Judgement of Taste*. Cambridge, MA: Harvard University Press.
Bowes, S. and Dawson, C.P. (1999) Watercraft user motivations, perceptions of problems and preferences for management actions: Comparisons between three levels of past experience. In H.G. Vogelsong (ed.) *Proceedings of the 1998 Northeastern Recreation Research Symposium – Administration/Education/Policy* (pp. 149–55). US Forest Service, General Technical Report NE 255.
Bristow, R.S., Klar, L.R. and Warnick, R.B. (1992) Activity packages in Massachusetts: An exploratory analysis. *Proceedings of the 1992 Northeastern Recreation Research Symposium* (pp. 33–6). US Forest Service, General Technical Report NE 176.
Cordell, H.K., McDonald, B.L., Briggs, J.A., Teasley, R.J., Bergstrom, J. and Mou, S.H. (1997a) *Emerging Markets for Outdoor Recreation in the United States*. Participation –Trends – Profiles (pp. 58–69). Athens: Sporting Goods Manufacturers Association and the USDA Forest Service.
Cordell, H.K., Tealsey, J., Super, G., Bergstrom, J.C. and McDonald, B. (1997b) *Outdoor Recreation in the United States. Results from the National Survey on Recreation and The Environment*. Available at www.ffs.fed.us./rvur...tions/outdoor_recreation/title.htm. Accessed 20 February 1999.
Cordell, H.K., McDonald, B.L., Tinsley, R.J., Bergstrom, J.C., Martin, J., Bason, J. and Leeworthy, V.R. (1999) Outdoor recreation participation trends. In H.K. Cordell (ed.) *Outdoor Recreation in American Life: A National Assessment of Demand and Supply Trends* (pp. 219–321). Champaign, IL: Sagamore Publishing.
Cordell, K, Green, G., Carter, B., Fly, M. and Stephens B. (2004) Recreation statistics update. Update report No 2. August. Available at www.srs.fs.usda.gov/trends/recupdate0907.pdf. Accessed on 14 October 2004.
Ditton, R.B., Goodale, T.L. and Johnsen, P.K. (1975) A cluster analysis of activity, frequency and environment variables to identify water-based recreation types. *Journal of Leisure Research* 7 (4), 282–95.
Driver, B.L., Tinsley, H. and Manfredo, M. (1991) The paragraphs about leisure and recreation experience preference scales: Results from two inventories designed to assess the breadth of the perceived psychological benefits of leisure. In B.L. Driver, P.J. Brown and G.L. Peterson (eds) *Benefits of Leisure* (pp. 263–86). State College: Venture Publishing.
DuWors, E., Reid, R., Bouchard, P. Legg, D., Boxall, P. Williamson, T., Bath, A. and Meis, S. (1999) *The Importance of Nature to Canadians: Survey Highlights*. Ottawa: Environment Canada.
Elder, G.H. Jr (1975) Age differentiation and the life course. *Annual Review of Sociology* 1, 165–90.
Green, G.T., Cordell K. and Stephens, B. (2003) Boating trends and the significance of demographic change. Presentation at the International Boating and Water Safety Summit in Las Vegas, NV, 15–17 April. Available at www.srs.fs.usda.gov/trends/nasblalv.htm. Accessed 23 September 2004.
Greenacre, M.J. (1984) *Theory and Applications of Correspondence Analysis*. Westport: Academic Press.
Hendee, J.C., Gale, R.P. and Catton Jr W.R.A. (1971) Typology of outdoor recreation activity preferences. *Journal of Environmental Education* 3 (1), 28–34.

Kelly, J. (1974) Socialization toward leisure: A developmental approach. *Journal of Leisure Research* 6 (3), 181–93.

Liikkanen, M., Pääkkönen, H., Toikka A. ja Hyytiäinen, P. (1993) Vapaa-aika numeroina 4. Liikunta, ulkoilu, järjestö- ja muu osallistuminen, loma, huvit. (Free time in numbers. Participation in sports, outdoor recreation, voluntary organisations and other leisure activities, vacation and entertainment) Tilastokeskus (Statistics Finland) SVT (Official Statistics in Finland), *Kulttuuri ja viestintä* (*Culture and the Media*), 6.

Lime, D.W. (1986) River recreation and natural resource management: A focus on river running and boating. President's Commission on Americans Outdoors: A Literature Review. *Management* 137–50. Washington, DC. The President's Commission on Americans Outdoors.

London, M., Crandall, R. and Fitzgibbons, D. (1977) The psychological structure of leisure: Activities, needs and people. *Journal of Leisure Research* 9 (4), 252–63.

McCool, S.F. (1978) Recreation activity packages at water-based resources. *Leisure Sciences* 1 (2), 163–73.

Pouta, E. and Sievänen, T. (2001) Ulkoilutilastot (Outdoor recreation statistics). In T. Sievänen (ed.) *Luonnon virkistyskäyttö 2000* (*Nature-based recreation 2000*). Metsäntutkimuslaitoksen tiedonantoja (Finnish Forest Research Institute, Research Papers) 802, 207–335.

Rapoport, R. and Rapoport, R.N. (1975) *Leisure and the Family Life Course* (p. 396). London and Boston, MA: Routledge & Kegan Paul.

Roos, J-P. (1987) Suomalainen elämä. (The Finnish Life). *Suomalaisen kirjallisuusseuran toimituksia*, 454. Hämeenlinna: Karisto Oy.

Sievänen,T., deVries, S., Scrintzi, G. and Floris, A. (2000) The recreational functions of European forests. In B. Krishnapilay *et al.* (eds) *Forests and Society: The Role of Research*. Vol. 1 (pp. 453–63). Proceedings of XXI IUFRO World Congress 2000, 7–12 August 2000, Kuala Lumpur.

Sievänen, T. (ed.) (2001) *Luonnon virkistyskäyttö 2000* (*Nature-based recreation 2000*). Metsäntutkimuslaitoksen tiedonantoja (Finnish Forest Research Institute, Research Papers) 802, 204.

Sievänen, T. Neuvonen, M. and Pouta, E. (2003) Veneilijöiden harrastajaprofiilit. (Boater groups and their profiles). *Liikunta ja Tiede* 5–6, 44–51.

Statistics Netherlands (CBS) (1997) Dagrecreatie 1995/96 (Daytrips 1995/96). Voorburg/Heerlen. The Netherlands: Statistcs Netherlands.

Tisley, H.E.A. and Johnson, T.L. (1984) A preliminary taxonomy of leisure activities. *Journal of Leisure Research* 16 (3), 234–44.

Veal, A.J. (1993) The concept of lifestyle: A review. *Leisure Studies* 12, 233–52.

Virtanen, V., Pouta, E., Sievänen, T. ja Laaksonen, S. (2001) Luonnon virkistyskäytön kysyntätutkimuksen aineistot ja menetelmät. (The data and methods used in a national survey of nature-based recreation demand). T. Sievänen (eds.) *Luonnon virkistyskäyttö 2000* (*Nature-based recreation 2000*). Metsäntutkimuslaitoksen tiedonantoja (Finnish Forest Research Institute, Research Papers) 802, 19–31.

Part 4: Planning and Management Issues

Chapter 10
Planning and Management of Lake Destination Development: Lake Gateways in Minnesota

WILLIAM C. GARTNER

Introduction

Many principles of tourism development have their roots in rural tourism development that includes lake-rich areas. Even though there is no universally accepted definition for rural tourism, one of its key features is that it takes place in peripheral areas – i.e. away from urban centres. Early tourism development work was almost exclusively focused on the rural setting. The origins of this work can be traced to the late 1960s and early 1970s. Jafari (1988) refers to this period as the 'advocacy platform'. During this time most of the published works on tourism were supportive of the activity, often touting its beneficial (mostly economic) impacts. Almost all the articles espousing the benefits of tourism development during this formative scholastic period refer to examples of tourism development that take place in rural areas. Gunn (1979) in one of the earliest books on tourism development, offered a view of tourism development that included attractions, transportation linkages and gateways or service centres. The gateway communities were viewed as a place to obtain needed services (e.g. lodging and food) before travelling, usually by road, to access the main purpose of the trip – the attraction. By contrast, an urban tourism development model would view the gateway and the attraction as inseparable.

In the mid-1970s, numerous academic studies were published that countered many of the favourable benefits of tourism development espoused by adherents of the advocacy platform. This plethora of studies has been labelled the 'cautionary platform' (Jafari 1988). It is during this time that rural-based tourism development studies began to assess some of the environmental and sociocultural impacts resulting from unplanned or poorly planned developments. As a result of the criticism, new community development models began to appear. Terms such as eco-tourism, cultural tourism, and green tourism began to appear which, in some cases, became major selling themes for the tourist trade. Studies proposing new models of tourism development were categorised as part of the Adaptancy platform (Jafari 1988). Most of the new models called for less intrusive types of development, more sensitivity to local needs and a greater reliance on local capital for development. Since urban areas were already physically transformed most of the attention for these new types of tourism development was centred on rural or peripheral areas.

Demand for the touristic use of rural areas has accelerated in recent years. Qualities inherent in a rural setting, such as personal contact, authenticity, heritage and individualism, resonate with an increasingly urban-based population (Long & Lane 2000). Media attention on the 'authenticity' of rural areas and, especially in the United States, a rural life that some see threatened by the expansion of large retailers (e.g. Wal-Mart), global food-service chains (e.g. McDonald's) and the loss of a traditional rural economic base (i.e. agriculture) has led to the search for the 'unspoiled' rural community. Unfortunately, it is exactly the unspoiled nature of the experience that may result in the rapid transformation of the resource base to accommodate increasing amounts of visitors. Understanding and exploiting tourism for rural communities while trying to maintain a traditional lifestyle is a difficult process (Perry et al. 1986; OECD 1994).

This chapter will examine some of the necessary ingredients for successful rural tourism development, with a special emphasis on lake destinations. It will not delve too deeply into many of the problems experienced by community residents as a result of tourism development. Instead, the reader is referred to Van der Stoep (2000), Gartner (1996) and Stokowski (2000) for more depth on this subject. Some of the key ingredients for opening up rural areas to tourists will be examined here. It is recognised that this approach can lead to many of the problems cited by the authors who have studied tourism's impacts on host communities. However, failure to understand the development process can lead to even more serious consequences if development proceeds without proper planning.

What is Rural?

Defining rural may seem to be an elementary exercise. Countries are fairly good at keeping accurate numbers of where people are living. Using a standard definition of the amount of people residing in a particular area is one criterion that is fairly commonly used to categorise communities as urban or rural. In the United States, the Economic Research Service (ERS) classifies any community with less than 2500 permanent residents as rural. Communities with 2500 to 19,999 permanent residents are classified as less urbanised, with any community over 20,000 classified as urban. The US Census Bureau defines rural areas as all non-urban areas, with urban defined as a community of over 50,000 permanent residents. Other federal agencies use other definitions for what is rural. For a more in-depth review of the various classification schemes currently used to define rural, the reader is referred to Flora *et al.* (1992). All the definitions cited above have a key element – i.e. the number of people who live within the area being observed. What is interesting is that many rural areas have peak seasons where the number of visitors greatly exceeds the number of permanent residents. During these peak periods the number of people found within the rural area may exceed the numerical definition of what makes an area rural.

Long (1998, in Gartner 2000: 301) proposes a definition of rural that reflects lifestyles one is likely to encounter in a visit to a 'rural' community: 'Rural can be perceived as a place of safety, with solid values, surrounded by open space and natural beauty, where one is treated respectfully and friendly.' In a functional sense, rural can be considered a place where small-scale enterprises dominate the economic scene; open space is abundant, contact with nature or 'traditional societies' is offered, development is slow-growing, using local capital, and the types of touristic activity offered varies but reflects local resource capabilities (Lane 1994). Getz and Page (1997) argue that even local enterprises are capable of growing quickly and rural tourism is still a possibility, even with rapid transformation of the physical plant.

Lake tourism is rural tourism. The use of lakes for fishing, boating and shoreline home development are types of recreation and tourism associated with the rural tourism experience. Referring to Gunn's (1979) original schematic of tourism development, there must be some type of gateway or service centre for lake tourism to flourish. The development of these gateways is subject to some special considerations of the tourism product and trends that affect how we access and utilise these products.

Consumption Centres

One of the key trends of the latter stages of the 20th century and that continues unabated today concerns the development of consumption centres (Gartner & Lime 2000). Consumption centres work by removing inefficiencies in the tourist distribution system and by concentrating tourists in a relatively small area. Tourism products require different distribution systems than for conventional trade goods. The most important differences are:

(1) An inverted distribution channel. Instead of a product flowing from manufacturer to distributor to consumer, the tourist must come to the source of production.

(2) Tourism products are produced and consumed at the same time. There is no shelf-life for the tourism product and all consumption is individually unique.
(3) Tourism, as a luxury good and one that can not be pretested, is heavily dependent on external forces, such as regional conflicts, internal political problems or health issues, all of which can have a major impact on use in the short term.

Because of the product differences noted above, the tourism trade has developed an extensive distribution system for moving large amounts of tourists. Given where the product is located, there may be as many as five separate business entities engaged in the distribution channel. A typical arrangement for a tourist wishing to travel to a remote region of the world may look like that depicted in Figure 10.1.

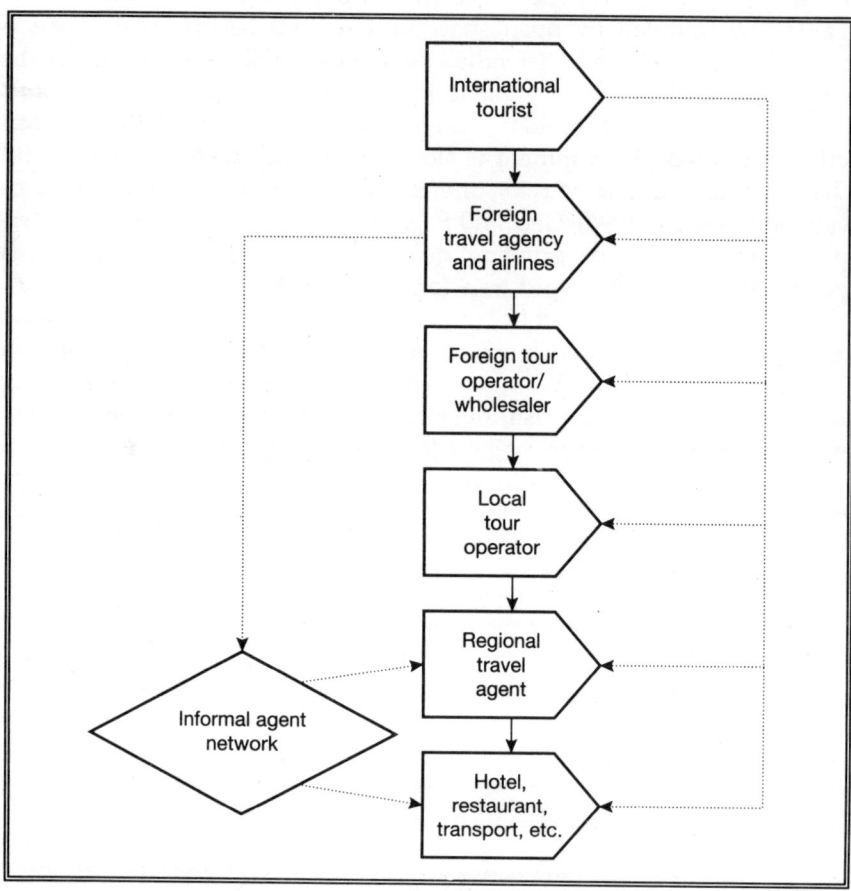

Figure 10.1 International tourism distribution channel

Each business existing along the distribution channel must make a profit to survive. Therefore, by the time the tourist buys the package, each individual unit (e.g. lodging) may cost much more than its original price to the first middle agent. Obviously, this potentially raises costs to the consumer and reduces revenue for the product provider, as much of the 'profit' ends up as middle-man revenue. One of the key selling points for packaged tours is their ability to deliver a number of tour components together at a price lower than could be purchased separately by a single consumer (Cavlek 2000). However, the existence of the 'jala gringo' component in Figure 10.1 refutes that point for certain package tours. 'Jala' meaning 'pull' and 'gringo' meaning 'foreigner' is the name given to individuals who meet arriving international tourists and offer to sell them tour components at 'local' prices. They do this by entering into secret deals with local product and service providers, thereby eliminating as many as three levels of the distribution channel.

Consumption centre development is also an attempt to eliminate the inefficiencies encountered along the distribution channel by increasing tourist numbers to a central location and in the process fostering competition among firms vying to bring tourists to the destination area. As consumption centres develop, they look for ways to prevent any firm from controlling access to the destination. First they do this by developing an area with high demand that leads to increased competition from the firms that control access to the destination. Second, consumption centres offer an extensive array of tourism products in a central location, thereby eliminating or sharply reducing the need for transportation linkages between the attractions and the gateway. A highly successful consumption centre blurs the lines between attraction and gateway. Some of the most successful consumption centres in the United States include Las Vegas, Nevada; Branson, Missouri; and Orlando, Florida. All three of these areas were at one time considered rural outposts, but all have made the transformation to urban-based consumption centres.

Gateways to lake destinations may also be viewed as consumption centres. The main difference between those mentioned above and those to be described in the next section is that it is not inevitable that consumption centre development leads to a transformation from a rural centre to one with urban characteristics, although there is a tendency to move in that direction. Reid and Smith (1997) building on the work of Gunn (1979) list six features that are key to successful tourism development for gateways: a sense of place; product/market match; clustering; transportation linkages; partnerships; and environmental protection. Of the six features, product/market match will exert the most influence on level of development. The examples below will help to illustrate this point.

Lake Gateways in Minnesota

Minnesota calls itself the land of 10,000 lakes. There are actually more than 10,000 lakes in the state, but the land of 11,081 lakes just does not have the same ring to it. Minnesota brands itself as a lake destination by using the 10,000 lakes slogan. The image of Minnesota as a outdoor recreation state tied to water was reinforced in the only image study done for the state (Nadkarni & Gartner 1989). As a result, one of the principles of image formation, 'the smaller the entity is in relation to the whole the less a chance to develop a separate image from the larger entity' (Gartner 1996) comes into play. Even though lakes are not numerous throughout the entire state but exist in abundance in only a number of areas, the dominant image of Minnesota is of lakes. The draw of lake gateway destinations is naturally tied to the lake resources within its visitor catchment area and these resources ultimately determine market demand. To better illustrate that argument, selective visitor profile data from three northern Minnesota lake gateways is presented (Gartner *et al.* 2001). The three gateways are Detroit Lakes, Brainerd Lakes area and Ely.

Detroit Lakes is located in the north-west central part of the state. It is 3½ hours away from the cities of Minneapolis and St Paul (Twin Cities) which have a population of over three million people, but are only 45 minutes from Fargo, North Dakota, a city of approximately 90,000 people. Primary activity in the area is tied to lake recreation such as swimming and fishing. A number of resorts, state parks and wildlife areas are found in the vicinity surrounding the city of Detroit Lakes. The area was once prairie which has now been converted to agricultural use. Two major outdoor concerts are held during the year at a separate concert centre near the city.

The Brainerd Lakes area consists of a number of smaller communities, with the city of Brainerd originally being the first point of contact for visitors from the southern part of the state where the vast majority of them originate. The region is located in central Minnesota in one of the state's premier forest/lake areas. When modern highway transportation systems were developed shortly after the Second World War, the Brainerd Lakes area became a major destination for resort clients and seasonal home owners. It is approximately 2½ hours from the urban centres of the twin cities.

Ely is located in the north-east part of the state, approximately 4½ hours from the twin cities. Of the three gateway communities described in this section, Ely looks and feels more like a tourist town as the majority of its retail businesses along the main street offer products and services in demand by tourists. It is also the main gateway to visitors venturing into the Boundary Waters Canoe Wilderness area (BWCWA), a federally designated wilderness managed by the US Forest Service.

Each of the above gateway communities owes it tourist economy to the type of resource available for use and the location of its primary markets – in other words, the product/market match. Community comparisons based on visitor profile analysis should help to reinforce this point.

Visitor Profile Analysis

Minnesota is considered a 'drive-to' state. The catchment area for Minnesota tourism is the 13-state north central region, of which Minnesota is a part, in the Midwestern United States. Few tourists fly into the state and, when they do, the primary reason is to visit friends and family (VFR) and/or attend conventions. This is especially noted for the small international market Minnesota claims (Gartner, *et al*. 2000). Even though research has revealed an 'average' visitor to Minnesota (TIA 1999), in reality this average visitor does not exist, especially when visitor profiles are examined. Differences between visitors using the different gateways tells us a great deal about the type and level of development one finds at the gateway.

Planning Horizon

Visitors to each of the three gateways exhibit different planning horizons. Brainerd is the shortest at approximately 70 days, followed by Detroit Lakes at 81 days and Ely at approximately 110 days (Figure 10.2). The reason Ely has such a long planning horizon is due to the restricted permit process for overnight stays in the BWCWA. Those wishing to venture into the BWCWA for one or more overnights must obtain an entry permit for the May to September period. Permits are controlled and available on a first-come first-served basis, beginning 1 January of each area. Brainerd has the shortest planning horizon of the three gateways featured. This is most likely due to the high percentage of seasonal home-owners found in the area. Seasonal home-owners do not have to worry about making reservations to stay in their house.

Average expenditure and type of accommodation

Average per party expenditure is highest for the gateway of Ely, followed by Brainerd and Detroit Lakes (Table 10.1). Part of this difference is due to the type of activities available in each area. Ely, being the gateway to the BWCWA, has many outfitters who rent all the equipment needed to stay in the wilderness for short to long periods of time. Brainerd, as will be shown, draws most of its visitors from the twin cities and, with the large seasonal home market in the area, many of the shops are the same or similar to those found in the twin cities. Detroit

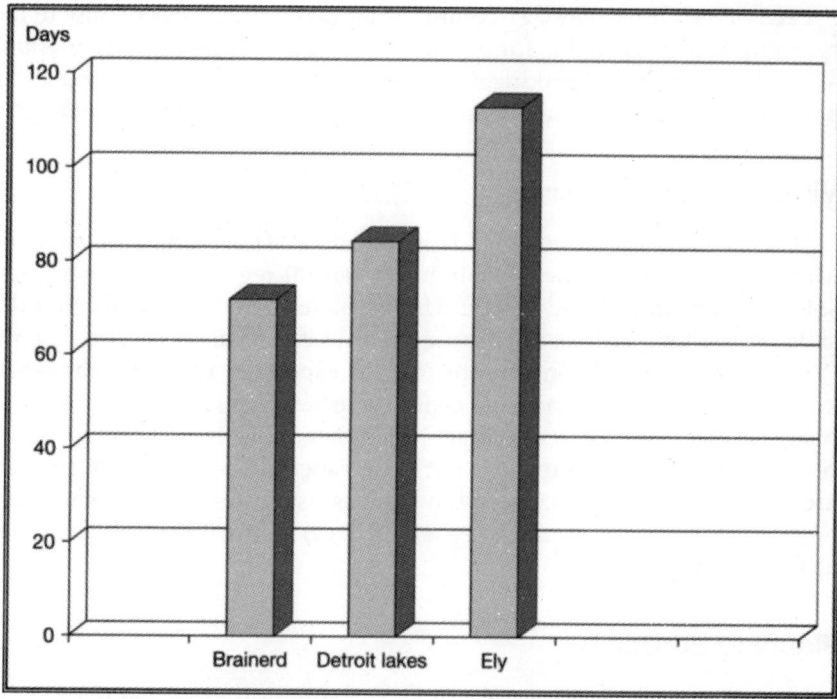

Figure 10.2 Planning horizon

Lakes has fewer shops that catering for tourists than either of the other two cities, as tourism is not its most important economic activity. It also does not have as large a second-home market as Brainerd, so fewer sales of the type required by seasonal home-owners are recorded. This is supported by the lower average shopping and grocery expenses recorded for Detroit Lakes tourists versus those visiting either Brainerd or Ely. Note, however, that accommodation in Detroit Lakes is more expensive than either of the other two gateways. This reflects a higher rate of stay at commercial lodging establishments for visitors to Detroit Lakes versus the other gateways. Visitors to Brainerd record higher levels of stay at seasonal homes or with friends and relatives, and Ely records higher levels of stay at resorts or commercial accommodation than for the other two gateways (Figure 10.3).

Prior visitation and primary residence

Brainerd records the highest level of prior visitation of the three gateways. Over 90% of Brainerd's tourists have visited, before compared to

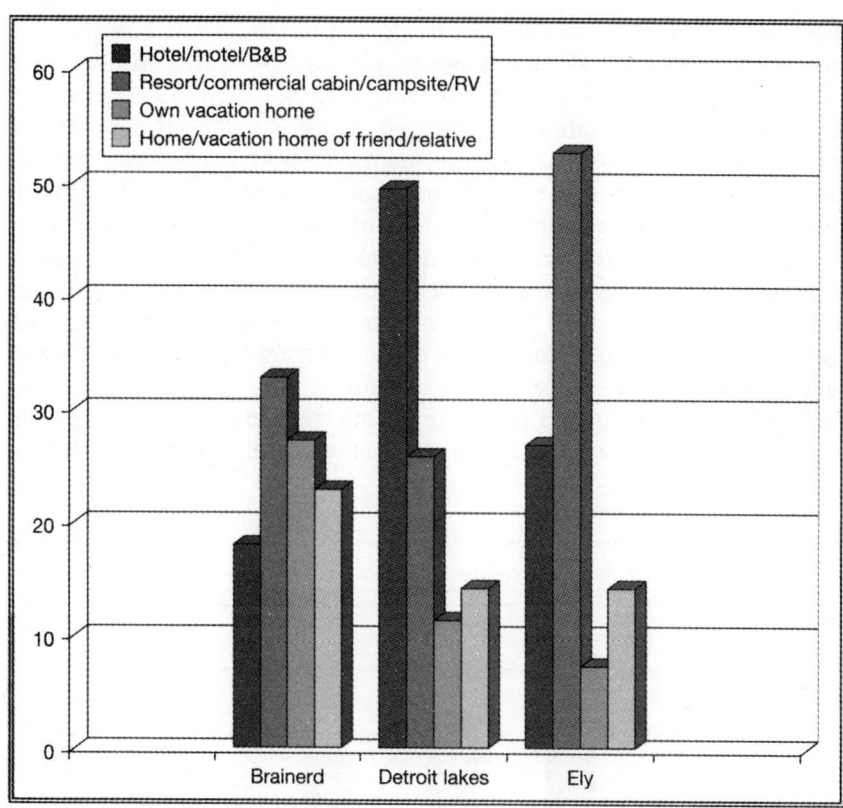

Figure 10.3 Types of accommodation

Table 10.1 Average daily expenditure per party

Spending categories	Brainerd US$	Detroit Lakes US$	Ely US$
Lodging	90.29	120.92	104.02
Restaurants/bars	65.23	43.56	55.44
Transportation	23.95	21.42	30.72
Groceries	22.92	13.13	21.40
Shopping	57.37	38.05	77.62
Miscellaneous	13.75	6.45	8.27
Total	305.22	253.17	329.75
People per party	3.84	3.55	3.63

approximately 81% of the visitors to Detroit Lakes and 74% of the visitors to Ely (Figure 10.4). This is not surprising, given that Minnesota is a 'drive-to' state and within easy driving distance of millions of people. However, when the state of primary residence is analysed over 90% of the visitors to Brainerd are Minnesotans. Detroit Lakes only records slightly over 50% of it visitors as originating in Minnesota, but this is misleading since the majority of the out-of-state visitors are from Fargo, North Dakota, a border city. Detroit Lakes is the nearest resort area, with significant water resources for residents of Fargo. Fargo residents are much closer to Detroit Lakes than those from the twin cities, so with regard to the average distance between market and destination, Detroit Lakes actually has a smaller visitor catchment area. Ely records approximately 55% of its visitors from Minnesota, with the rest coming from a wide variety of markets (Figure 10.5). Ely has the widest catchment

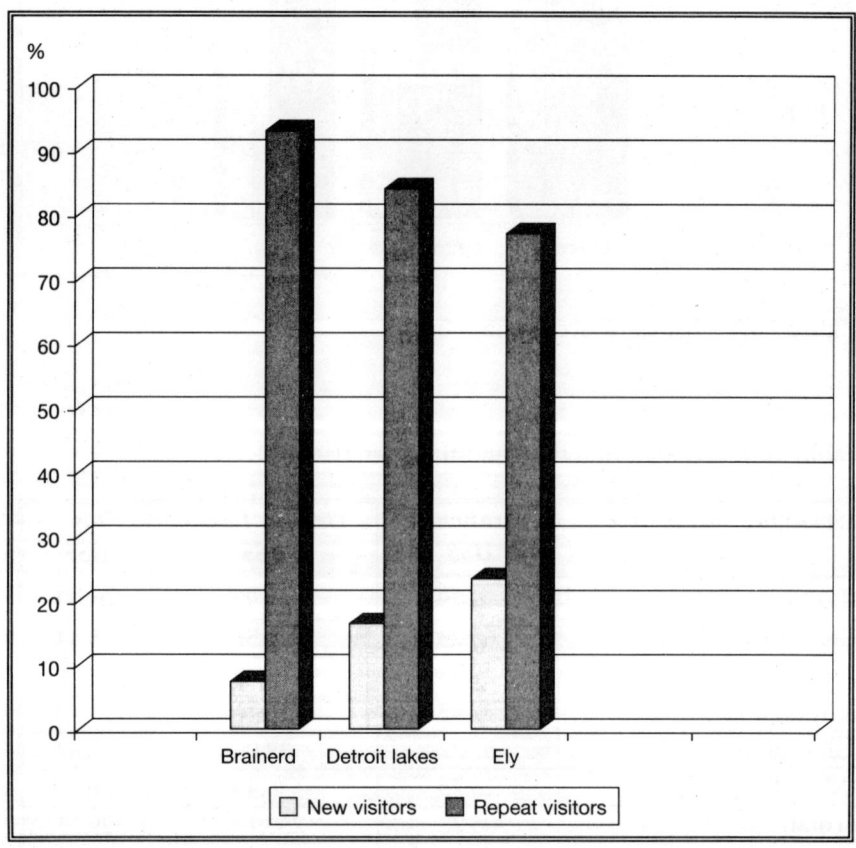

Figure 10.4 Prior visitation

area, as it is the only community outside of the twin cities to actually record significant numbers of international visitors. Ely's draw is almost entirely due to the wilderness resources next to it and the experiences to be gained from visiting this protected area and enjoying its facilities.

Discussion

The above discussion reveals that not all rural gateways will develop in a similar manner. How they develop is first and foremost tied to the resources available for use. In the above example, Detroit Lakes developed not necessarily for tourism but has incorporated tourism, into its overall development scheme. Visitors, even though many come from

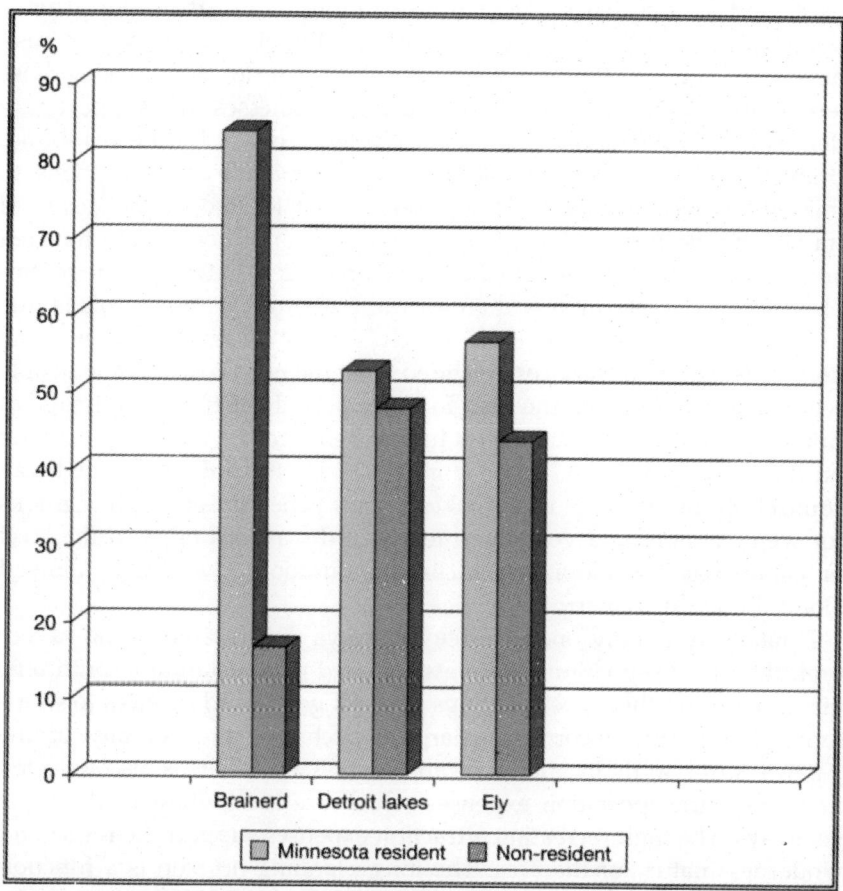

Figure 10.5 State of primary residence

outside the state, do not travel very far to reach Detroit Lakes and generally do not stay in holiday homes. As a result, expenses are lower for items such as transportation, restaurants and bars, groceries and shopping. Most visitors will stay at some form of commercial lodging establishment and therefore do not need to shop for many household items in the destination community. Another factor is that much of the commercial lake-based accommodation is away from Detroit Lakes in a scattered and a dispersed setting. Detroit Lakes still functions as a gateway community, but its overall importance as a gateway is lessened by how the resources around it are used by visitors.

Brainerd, on the other hand, records high expenditure for items such as groceries, shopping and restaurants/bars. Many people, when visiting the Brainerd Lakes area, stay at their own, or a friend's, holiday or permanent home. It is also the primary resort area for Minnesota residents and as such has developed much more of an urban look. This is not only noticeable by expenditure patterns, but also by the type of businesses operating in the area which are similar, if not the same, as those left behind in the twin cities. The types of businesses that support residential developments in urban areas are the types of businesses found, primarily, in the Brainerd Lakes area. There are some exceptions to this, but tourist companies are generally not located at the nexus of traffic activity, but instead are to found at the northern, more forested areas of the region. Brainerd has developed as a gateway that resembles more of an urban area than a rural tourist area. Again, this is due to the type of attraction on offer, which is heavy on resorts and seasonal homes. The types of services provided meet the needs of the same people who have been visiting the area for long periods of time. The image of Brainerd as an urban gateway is further reinforced by a study of users on the main corridor (Gartner *et al.* 2002). Users of the main highway connecting the gateway to the lakes region generally give it low marks for scenic quality and voice high levels of dissatisfaction with the level of commercial development, including outdoor advertising, allowed along the highway corridor.

Finally, Ely is only one of three gateways that resembles the stereotypical tourist destination. Its streets are lined with restaurants, outfitters, gift shops and other types of shops that sell goods and services to short-term visitors. Ely records the largest catchment area of any of the gateways due to its location next to the BWCWA. This is also reflected in average transportation expenses which are the highest of the three gateways. The high recreation attraction expenses support the resort and wilderness nature of the area. The long planning horizon is a function of a permit system that controls visitor use in the BWCWA.

Conclusions

The above three destination examples are used to show how level and type of gateway development is tied to the resource base and how that resource is utilised. Here, the key elements, identified by Reid and Smith (1997), of sense of place, clustering, environmental protection, partnerships and transportation linkages come into play. Transportation because of the time and cost associated with it is a key element in the destination development process. The longer it takes and the more expensive it is to reach a destination, the higher the perceived quality of the product in place. How resources are designated, from an environmental perspective, determines the perceived resource quality, which affects drawing power. Clustering is a policy tool that can make a destination appear a certain way. In Brainerd, for example, clustering of services was not employed but rather businesses were left to develop on their own, in whatever way they chose. How they have developed, along a main transportation route into the destination, is generally perceived in a poor light by the people using the route. Finally, sense of place is an amalgam of the development that has resulted, whether planned or unplanned, from the product that is in place and the market that has been attracted to the product. Gateway communities, both old and new, really do control their own destiny, but many residents of these destinations may not realise how to make the system work for them. The types of tourist development encouraged can lead to certain firms developing. For example, if the lake resources around Ely were not protected from development by being in federal ownership and managed as a wilderness area, it is likely that far more holiday homes would be found there. If that were the case, more shops that cater to a home market rather than a transient tourist market would be encountered. Instead, the type of use allowed, which can also be viewed as the product, does a great deal to encourage particular types of business development.

Consumption centre development is a trend that has accelerated in recent years. In a way, it is a model that takes the gateway approach first proposed by Gunn and incorporates all the attractions into a central area, thereby reducing the need for transportation linkages. Rural areas that adopt this approach then begin to transform into something resembling an urban tourism destination. Some major tourist destinations that in the off-season truly are rural areas, become more urban-looking as tourists numbers increase during the season. However, it is not inevitable that all rural tourist destinations will be completely transformed into what resembles an urban destination. Rural lake areas, because of the dispersed nature of lake-based tourism will retain their rural setting, but depending on the type of development allowed on the lakes, the gateway may transform into something much more urban in nature. In truth, it may not actually be the level and type of development within the gateway that matters, but instead the degree of transformation that takes place where the primary visitor experience is achieved. In that sense, the future of the area will be tied to how the lake resources are perceived by those using them, with the level of development in the gateway being of secondary importance.

The planning and management of lake destination areas depends initially on the type of use that will be allowed with and around, the lake area. Once the type and level of use is determined, a product/market match can be examined. This will allow the analyst to predict fairly accurately the type and amount of visitors attracted to the area. Second, knowing who is coming will help to determine the types of businesses most likely to appear to service lake region visitors. Understanding and knowing what are the most likely development forces operating in an area is crucial for careful planning and management of the resource, and the gateway services and access to the lakes region. Patterns of development are fairly predictable, provided some advance information is known about the experiences that can be gained from travelling to the area. Using all the tools available to regional planners today, once development potential of an area is understood, management and planning can be undertaken with a fairly complete knowledge base of what type of development is most likely to occur. It is a given that destination transformation will occur. The degree and form that transformation takes is predictable and manageable. This means that sustainable development, a desirable concept although one that possesses great ambiguity, can be achieved. Although sustainable tourism development has not been discussed in this chapter, it is safe to assume that once the level and type of future development is known, there is a much better chance of meeting sustainable objectives, when they have been determined, for the lakes area in question.

References

Cavlek, N. (2000) The role of tour operators in the travel distribution system. In W.C. Gartner and D.W. Lime (eds) *Trends in Recreation, Leisure and Tourism*, Wallingford: CABI.

Flora, J., Spears, L., Flora, L. and Weinberg, M. (1992) *Rural Communities: Legacy and Change*. Boulder, CI: Westview Press.

Gartner, W.C. (1996) *Tourism Development: Principles, Processes and Policies*. New York: John Wiley.

Gartner, W.C. (2000) Rural tourism development. In W.C. Gartner and D.W. Lime (eds) *Trends in Outdoor Recreation, Leisure, and Tourism*, Wallingford: CABI.

Gartner, W.C. and Lime, D. (2000) The big picture: A synopsis of contributions. In W.C. Gartner and D.W. Lime (eds) *Trends in Recreation, Leisure and Tourism*, Wallingford: CABI.

Gartner, W.C., Limback, L. and Erkkila, D. (2000) *Barriers to Increasing Minnesota's Share of the International Visitor Market*. St Paul: Tourism Centre, University of Minnesota (www.tourism.umn.edu/research).

Gartner, W.C., Love, L. and Erkkila, D. (2001) *Study of Current Area Tourists; Visitor Profiles (Parts I and II)*. St Paul: Tourism Centre, University of Minnesota (www.tourism.umn.edu/research).

Gartner, W.C., Love, L. and Erkkila, D. (2002) *Attributes and Amenities of Minnesota's Highway System that are Important to Tourists*. St Paul: Tourism Centre, University of Minnesota (www.tourism.umn.edu/research).

Getz, D. and Page, S. (1997) Conclusions and implications for rural business development. In S. Page and D. Getz (eds) *The Business of Rural Tourism: International Perspectives*. London: International Thompson Business Press.

Gunn, C. (1979) *Tourism Planning*. New York: Crane Russak.

Jafari, J. (1988) Retrospective and prospective views on tourism as a field of study. Paper presented at the 1988 meeting of the Academy of Leisure Sciences, Indianapolis Indiana.

Lane, B. (1994) What is rural tourism. *Journal of Sustainable Tourism* 2, 7–21.

Long, P. (1998) *Rural Tourism Foundation Information Piece*. Boulder, CO: University of Colorado.

Long, P. and Lane, B. (2000) Rural tourism development. In W.C. Gartner and D.W. Lime (eds) *Trends in Recreation, Leisure and Tourism*. Wallingford: CABI.

Nadkarni, N. and Gartner, W. (1989) *Minnesota Image Study*. Menomonie: Centre for Hospitality and Tourism Research, University of Wisconsin-Stout.

Organisation for Economic Cooperation and Development (OECD) (1994) *Tourism Strategies and Rural Development in Tourism Policy and International Tourism*. Paris: OECD.

Perry, R., Dean, K. and Brown, B. (1986) *Counterurbanisation: International Case Studies*. Norwich: Geo Books.

Reid, L. and Smith, S. (1997) *Keys to Successful Tourism Development: Lessons from Niagara*. St Catharines, Ontario: Brock University (video and learning guide).

Stowkowski, P. (2000) Assessing social impacts of resource-based recreation and tourism. In W.C. Gartner and D.W. Lime (eds) *Trends in Recreation, Leisure and Tourism*. Wallingford: CABI.

Travel Industry Association of America (TIA) (1999) *Travelscope*. Washington, DC: TIA.

Van der Stoep, G. (2000) Community tourism development. In W.C. Gartner and D.W. Lime (eds) *Trends in Recreation, Leisure and Tourism*. Wallingford: CABI.

Chapter 11
Lake Tourism in New Zealand: Sustainable Management Issues

C. MICHAEL HALL AND MICHELLE STOFFELS

The notion of 'clean and green' is inextricably linked to the international perception and imaging of New Zealand tourism (Hall & Kearsley 2001). Crystal clear water, green mountains and valleys and pure white snow, along with the portrayal of traditional elements of Maori culture are all elements of the development of New Zealand tourism as '100% pure'. For example, a 2003 presentation from Tourism New Zealand, the country's international tourism marketing organisation, noted that what brings tourists to New Zealand are 'fiords and mountain ranges' among other natural and cultural products (Tourism New Zealand 2003a). In addition to the promotional slogan '100% pure', the advertising campaigns developed by Tourism New Zealand in recent years have used phrases such as '100% pure adventure', '100% pure solitude' and '100% pure adrenaline', and have featured images of 'iconic freshwater bodies including the Southern Lakes, the thermal waters of Rotorua and most recently, Lake Rotorua' (Ministry of Tourism 2004: 2). Arguably, such imaging strategies, complete with the use of lakes as a component of media promotion of New Zealand have only been reinforced by the association of New Zealand with such films as *Lord of the Rings* and the associated promotion of the country as 'Middle Earth'. Yet, with continued attempts to develop New Zealand's share of the international tourism market, just how clean and green can the country remain? The pressure is on New Zealand to 'preserve and promote natural, unspoilt environments' while simultaneously increasing income from international tourism (Tourism New Zealand 2003a). However, when we consider the freshwater resource of lakes in the country, perhaps preserving these pristine and populous areas is becoming increasingly difficult, given the tourism and recreational demands that are being placed upon them. As the Ministry of Tourism (2004: 1) states, 'New Zealand lakes and rivers feature prominently as tourism attractions and the sustainable development of these natural resources is critical to the current and future success of the tourism industry.' These issues provide the context for consideration of the role of lakes in New Zealand as tourist icons and attractions, and sites of extensive leisure, recreation and tourism activities.

As many international and domestic tourists in New Zealand base their holiday around 'natural wonders' (Tourism New Zealand 2003a), there is a clear need for serious consideration of the natural resources that support international tourists, as well as domestic tourists, recreationists and leisure seekers. While accurate longitudinal data is not available for domestic tourism, in 2001 New Zealand residents took 38.9 million day trips and 16.6 million overnight trips. Among the top 20 activities are beach walking, walking, swimming, fishing, sports and hot pools (Hall & Kearsley 2001; Tourism Research Council of New Zealand. As part of a programme to identify potential water bodies of national importance for recreation value, the Ministry for the Environment (MFE) (2004b) reported that 79% of New Zealanders identified themselves as recreational users of fresh water. Main uses identified in the study which examined rivers, lakes and wetlands were walking, general sightseeing, picnicking, swimming, fishing, whitebaiting and hunting (MFE 2004b: 11).

International tourist numbers reached 2 million arrivals in a 12 month period for the first time in 2001 (Tourism New Zealand 2003b). Of New Zealand's specific/unique attractions, geothermal sites attracted the greatest number of international visitors, followed by the visiting of beaches, many of which will be lake rather than sea beach experiences. As may be expected sightseeing is also a major activity. Table 11.1 indicates the lake- and lacustine-related activities and attractions undertaken by international visitors in New Zealand from 1997 to 2004 as estimated in the New Zealand international visitor survey. Unfortunately, with the majority of water-based activities, it is impossible to know whether they are salt or freshwater based. Nevertheless, lake fishing was estimated to attract over 33,000 international visitors in 2004, while lakes as a specified attraction drew over 80,000 international visitors. According to the Ministry of Tourism (2004: 24):

> international visitors who undertake freshwater activities in New Zealand generally stay longer than 'all visitors'. This is expected to some degree, as it is assumed that international visitors who have time to participate in adventure activities are more likely to visit New Zealand for a longer period of time.

The tourist 'hot-spots' of New Zealand occur throughout both the North and the South Islands and, apart from the obvious city destinations as tourist hubs, areas that follow close second as destination choices appear to rely on lakes as a component of their tourism resources. In the North Island, two of the main tourist areas are Rotorua, which is

Table 11.1 Lake- and lacustine-related activities and attractions undertaken by international visitors in New Zealand, 1997–2004

Activities/ attractions/reasons	Year end							
	Dec. 1997	Dec. 1998	Dec. 1999	Dec. 2000	Dec. 2001	Dec. 2002	Dec. 2003	Dec. 2004
Total (automatic base)	1,334,216	1,343,889	1,438,705	1,585,801	1,695,382	1,795,466	1,908,158	2,150,106
Beaches	77,219	325,352	294,734	305,281	371,931	372,197	555,155	653,670
Scenic boat cruise	263,670	319,234	288,862	349,318	298,205	320,399	440,032	590,917
Jet boating	143,271	172,899	158,308	178,636	174,820	170,639	200,414	224,599
Swimming	89,073	89,648	76,968	99,014	80,790	85,542	141,152	157,780
Fishing lake	29,238	36,345	36,454	33,166	38,836	31,766	41,774	33,738
Sailing	40,221	31,908	27,451	28,728	27,575	31,032	45,487	47,580
Fishing river	24,787	15,553	17,774	19,584	24,310	27,093	29,442	23,351
Kayaking river	14,292	13,626	10,802	22,074	18,791	24,547	25,494	26,625
Other water activities	4,012	19,874	16,423	25,485	17,840	22,554	45,459	40,981
Scuba diving/ snorkling	0	0	0	0	0	4,629	24,351	23,219

Table 11.1 Continued

Activities/attractions/reasons	Year end							
	Dec. 1997	Dec. 1998	Dec. 1999	Dec. 2000	Dec. 2001	Dec. 2002	Dec. 2003	Dec. 2004
Water skiing	0	13,526	3,780	3,901	2,482	4,255	7,735	5,310
Canoeing	0	0	0	0	0	0	0	1,421
Manapouri	0	0	0	0	0	0	0	92
Te Anau	0	0	0	0	0	0	0	1,718
Mt Aspiring/Lake Wanaka	0	0	0	0	0	0	0	3,360
Nelson Lakes	0	0	0	0	0	0	0	1,383
Lakes	0	0	0	0	0	0	0	80,626
Rivers	0	0	0	0	0	0	0	8,152

Note: Visitor counts less than 6000 are subject to greater than +/-50% error

Source: Derived from Tourism Research Council of New Zealand (2005) International Visitor Survey Activities/Attractions in NZ, www.trcnz.govt.nz/Surveys/International+Visitor+Survey/Data+and+Analysis/Table-Activities-Attractions-in-NZ.htm.

located beside Lake Rotorua and is adjacent to a significant lake complex which is a base for sightseeing, thermal pools, fishing and boating activities (see www.rotoruanz.co.nz/); and Lake Taupo, internationally renowned for its trout fishing since the early 1900s after the visit from author Zane Gray and his reporting of the wonderful fishery (see www.laketauponz.com/). In the South Island, there is New Zealand's premier alpine tourist resort, Queenstown, on the shores of Lake Wakatipu (see www.queenstown-nz.co.nz/); Wanaka, a town located close to a major ski field that has become a destination in its own right, based on the shores of Lake Wanaka (see www.lakewanaka.co.nz/); and Lake Te Anau, gateway to Milford Sound and World Heritage listed Fiordland National Park (see www.fiordland.org.nz/) (also see MFE 2004b; Ministry of Tourism 2004). Details of the largest lakes in New Zealand and their characteristics are listed in Table 11.2.

It is almost indeterminable how important these lakes are to the tourism industry, as well as the local population, although the Ministry of Tourism (2004) has identified some lakes and freshwater sources as being of significance in terms of their scenic value and as a site of tourism activities. Unfortunately, while the Ministry of Tourism's (2004) report notes the importance of freshwater resources for tourism activities as identified through international and domestic travel surveys, their data cannot be identified with specific lake locations in the majority of instances, although the Ministry did identify particular sites in terms of their significance. The MFE (2004b) study of freshwater bodies of national significance for recreation used a number of different methods to identify potential water bodies for inclusion in a final list, including:

Table 11.2 Ten largest lakes in New Zealand

Name	Island	Area (km^2)	Length (km)	Width (km)	Depth max (m)	Elevation
Taupo	North	622.6	40.5	29.5	162.8	357
Te Anau	South	352.0	60.0	28.6	417.0	203
Wakatipu	South	293.0	75.2	6.2	380.0	309
Wanaka	South	193.0	45.5	11.6	311.0	277
Ellesmere	South	180.0	26.3	12.9	4.0	1
Manapouri	South	153.0	28.3	11.5	444.0	179
Hawea	South	141.0	41.9	10.4	384.0	342
Pukaki	South	98.9	22.9	8.0	70.0	497
Tekapo	South	88.0	25.2	5.9	120.0	710
Wairarapa	North	80.7	18.2	9.6	2.5	2

- a random telephone and targeted Internet survey which led to the inclusion of water bodies identified by ten or more respondents;
- wetlands that were rated 'A' in a 1987 study of Wetlands of National Importance (Davis 1987);
- water bodies that were identified as having over 10,000 angler day visits in the 1994 or 2002 Fish and Game National Angling Surveys (Unwin & Brown 1998; Unwin & Image 2003);
- the most important whitebait fishery water bodies as identified by the Department of Conservation experts and representatives of whitebaiting associations;
- water bodies on which recreation-related conservation orders had been placed.

As a result of the amalgam of different types of significance, a list of 105 freshwater bodies was developed that were considered to be potentially of national importance for recreation. Of these 54 were rivers, 30 were lakes and 21 were wetlands (MFE 2004b: 6). The lakes that were identified as being of potential significance are listed in Table 11.3.

Despite the lack of systematic lake-specific research, lakes are clearly a major determinant in the location of permanent and second-home settlements, as well as recreation and tourist activities. If we are also to consider direct and indirect social benefits and economic benefits, such as hydroelectric power generation, water supplY and irrigation, the value to be quantified is undoubtedly considerable. Yet despite such values, there is a general lack of appreciation of the significance of lakes within New Zealand's economy and lifestyle, although recent government actions have attempted to focus on the conservation and management of New Zealand's freshwater resources (see the Water Programme of Action pp. 200–1). Instead, they are taken as a given, as a common resource THAT is regarded as being almost inexhaustible in terms of the demands made upon them. However, in recent years such perceptions have perhaps finally begun to change following not only the impact of droughts on some lakes and shore communities, but increasing problems with eutrophication and pollution.

Quality

When we conjure up images of fresh-water we think of clean and crisp clear water, or water bodies with beautiful hues of green and blue (James et al. 2002). We enjoy these pure images for recreational activities such as swimming and boating and for the simple replenishment of drinking water. However, if the state of that water quality is altered, or at least people's aesthetic perception of 'clean water' (House 1996) serious issues arise.

Table 11.3 Potential lakes of national importance for recreation, identified by the Ministry for the Environment

Water body	Region
North Island	
Kai Iwi Lakes	Northland
Pupuke Lake	Auckland
Taupo Lake	Waikato
Karapiro Lake	Waikato
Rotorua Lake	Bay of Plenty
Blue Lake	Bay of Plenty
Rotoiti Lake	Bay of Plenty
Tarawera Lake	Bay of Plenty
Green Lake	Bay of Plenty
Rotoma Lake	Bay of Plenty
Aniwhenua Lake	Bay of Plenty
Waikaremoana Lake	Hawke's Bay
Dune Lakes	Manawatu-Wanganui
Kahangatera and Kohangapiripiri lakes and wetlands	Wellington
Wairarapa Lake	Wellington
South Island	
Nelson Lakes	Tasman
Benmore Lake	Canterbury
Ellesmere Lake and wetlands	Canterbury
George Lake and Henderson extension	Canterbury
Aviemore Lake	Canterbury
Brunner Lake	West Coast
Okarito Lagoon	West Coast
Okura/Turnbull/ Hapuku Lagoon	West Coast
Tawharekiri Lakes complex Waita mouth	West Coast
Wakatipu Lake	Otago
Wanaka Lake	Otago
Hawea Lake	Otago
Dunstan Lake	Otago
Te Anau Lake	Southland
Manapouri Lake	Southland
Waituna lagoon and wetlands	Southland

Source: Derived from MFE 2004b

Aesthetically speaking, lake environments have much to offer. The aesthetic appeal includes the water quality, as well as flora and fauna in and around the lake. It is also paramount that enjoyment of the lake can occur without urban development encroaching on the lake and its views (Hall & Kearsley 2001) and agriculture and industry influencing the water and riparian habitats significantly. For example, in a report on the sustainability of intensive farming in the New Zealand environment the Parliamentary Commissioner for the Environment reported that:

> Water quality declines markedly in lowland streams and rivers in pasture-dominated catchments. Many rivers draining farmland are unsuitable for swimming because of faecal contamination from farm animals, poor water quality and nuisance algal growths caused by excess nutrients. (PCE 2004: also see Larned et al. 2005)

The issue of microbial contamination of New Zealand's lakes and rivers is substantial (Smith et al. 1996; Wilcock et al. 1999; Vant 2001). In a joint MFE and Ministry of Health (2002), study 60% of samples in the survey contained campylobacter and 59% viruses. The result of this and earlier studies indicated that 'concentrations of the faecal indicator E. coli often exceed 1000 colony forming units (cfu) per 100 ml, which far exceeds the E. coli guideline for fresh water recreation (median of 126 cfu/100ml)' (PCE 2004: 104). In fact, somewhat belying its clean green image, New Zealand has the highest reported incidence of campylobacteriosis in the developed world, with the annual economic cost of the disease estimated to be NZ$61.7 million in 1999 (Scott et al. 2000; PCE 2004). In addition, the extent of sedimentation on downstream lakes following land clearance is substantial (Ryan 1991; Mosley & Jowett 1999; Quinn & Stroud 2002), resulting in a number of impacts, including the degradation of substrates for bottom-dwelling organisms, reduced light penetration and visual clarity and infilling. It is estimated that the sediment yield in lakes and estuaries in New Zealand is about ten times greater than from the pre-existing native forest (Davies-Colley et al. 2003).

Although the state of groundwater quality is not comprehensively known at a national level, 'many shallow aquifers beneath dairying or horticultural land have elevated nitrate levels' (PCE 2004: 45). This situation has meant, for example, that in the case of Lake Taupo, there is a proposal for regulatory control of non-point source discharges into the lake as well as further advice, education and research on the development of low-nitrogen farm systems and land-use in order to reduce nitrogen discharge into the lake. The amount of nitrogen and phosphorus entering lake systems is significant because of their contribution to eutrophication and hypoxia. In New Zealand, plant growth in Lake Taupo and the lakes in the Rotorua region is limited in the absence of nitrogen. However the amount of nitrogen entering these lake systems has increased markedly in the past 50 years from both rural and urban sources (Rosen 2001). Nevertheless,

> nutrient enrichment of lakes from farming activities is a growing concern and is not only affecting shallow lakes but deeper lakes too, such as those in the Rotorua area. The lag time for nutrients to enter these lakes suggests that the problem will get worse before it gets better, even if measures are put in place to reduce nutrient inputs. (PCE 2004: 96)

One of the difficulties in determining the effects of water catchment practices on lake systems is that the 'cumulative impacts of groundwater pollution on farming practices may only become slowly evident over time' (PCE 2004: 24). Yet there have been indications that there are problems with water quality in New Zealand lakes in policy statements since at least the early 1990s. For example, the Ministry of Agriculture and Forestry (MAF) reported in 1993 that

> ... more than 700 lakes [in New Zealand] are shallow and between 10% and 40% of these are nutrient enriched (eutrophic). Most of the eutrophic lakes are in the North Island and in pasture dominated catchments. A number are subject to fish kills or are no longer capable of supporting fish life ... development of their catchments, primarily for agriculture, is almost certainly responsible, due to the substantially increased nutrient loads that result.

Management responses have been developed for some lakes in order to respond to the problems of nitrogen and phosphorus inputs and other factors that affect water quality. In the case of Lake Taupo, the 2020 Taupo-nui-a-Tia Action Plan aims to reduce the manageable sources of nitrogen flowing into the lake by 20% by the year 2020, with the Ministry for the Environment (MFE) (2004a: 1) recognising that 'Water quality helps underpin New Zealand's clean, green image. Trout fishing and the future of tourism in the Taupo region depend on the condition of the lake.'

The development of a strategy for the management of nutrients into Lake Taupo occurred in three distinct phases (Table 11.4). The 2020 project was especially important as it provided 'an overarching framework for the future resource management of Lake Taupo' (PCE 2004: 153). The project illustrates some of the elements of a more comprehensive approach to lacustrine system management, as suggested in Chapter 1. The project utilised a long-term perspective, incorporated a wide range of management issues into understanding lake and catchment management problems and involved significant levels of community participation. Issues that were identified as posing a threat to the lake included:

- increasing human use and activities, particularly with respect to recreation and tourism use;
- increased levels of nutrients entering the lake;
- increased levels of faecal organisms in lakes and waterways;
- the impacts of boating in terms of the introduction of weeds, littering, the discharge of waste, as well as increased competition between boats and other recreational activities such as swimming;
- changes in lake levels;
- perceived lack of effective management strategies by various organisational stakeholders and the lack of cooperation between those organisations.

The final phase in the development of a management strategy for the lake has been the development of a new planning framework for the protection of water quality in the lake, known as the Protecting Lake Taupo Project. The principal target of the planning framework is 'over the next 15 years, to reduce the manageable sources of nitrogen flowing into Lake taupo by 20%' (PCE 2004: 154). The planning framework designates certain actions that would reduce nitrogen run-off into the lake, as well as the regulatory instruments that underly them. The total cost

Table 11.4 Phases in the development of a strategy to manage nutrient inflow into Lake Taupo

Period	Phase
2000	Communication of information from Environment Waikato's Lake Taupo long-term water quality monitoring programme to various forums in the regional community that outlined the state of the catchment and the declining health of the lake.
2001–3	Debate and discussion within the regional community about the state of Lake Taupo, while agencies and the community also work out how to achieve the desired level of lake health. This phase was primarily managed through the 2020 Taupo-nui-a-Tia project which was a three-year Ministry for the Environment Sustainable Management Fund project focused on incorporating community values and aspirations (including those of the local iwi) into an action plan.
2004–	Development of a planning framework to ensure actions that are taken to protect Lake Taupo in the light of the findings and actions of the 2020 Taupo-nui-a-Tia project.

of the project is estimated to be NZ$81.5 million over 15 years, of which central government contributes approximately 40% (MFE 2004a). Actions under the plan include:

- establishment of a joint public fund derived from local and regional property taxes and national government funding to help convert pastoral land to low-nitrogen land in the most cost-effective way. Mechanisms include, land purchase and change of use (including forestry, public use, recreation and biodiversity), covenanting of land so that it can only be of low-nitrogen production, joint ventures and/or direct purchase of nitrogen discharge;
- using Regional Plan rules to restrict current levels of nitrogen being lost into the catchment;
- upgrading and better maintaining sewage systems;
- assisting the development of low-nitrogen farming practices;
- developing strong partnerships between the local *iwi* (Ngati Tuwharetoa) and government;
- exploring low-nitrogen land-use options, such as native forest planting and regeneration which would have significant biodiversity values, as well as 'low-impact tourism and recreation facilities. (PCE 2004: 154)

The Rotorua Lake system is also experiencing issues with respect to declining water quality. However, in this major tourism centre that attracts over half a million international visitors each year, 'the situation is even more critical' (PCE 2004: 100). In the case of the Rotorua lakes, nutrients have entered the lake system via farming practices and from septic tanks, many of which are from second homes and have reduced dissolved oxygen levels and triggered toxic blue-green algal blooms (Hamilton 2003). In response, a cooperative conservation and restoration strategy has been developed by local government agencies, including Environment Bay of Plenty and the Rotorua District Council and the Te Arawa Maori Trust Board. The vision for the lakes is:

> The lakes of the Rotorua district and their catchments are preserved and protected for the use and enjoyment of present and future generations, while recognising and providing for the traditional relationship of Te Arawa with their ancestral lakes. (Hamilton 2003: 26)

The strategy outlines a number of goals with respect to protection, use, enjoyment and management (Table 11.5) and has received NZ$7.2 million from the national government towards improving water quality (Hobbs 2004; PCE 2004). Interestingly, when the funding was announced, the Minister for the Environment, Marian Hobbs, noted the significance of tourism:

Rotorua is on an important tourist circuit with over half a million international visitors each year ... Lake Rotoiti is a key tourism asset but the water quality has been deteriorating over many years and it has become significantly worse in the last few years. We want to see it cleaned up.

The water quality issues at Lake Taupo and the Rotorua lakes highlights the importance of an ecosystem approach to lake management. Indeed, it is critical to recognise that any effective lake-management strategy must utilise a watershed or river basin framework so that all inputs into the lake system can be addressed. Unfortunately, in the New Zealand situation there exists a lack of comprehensive knowledge as to the effects of human activities on lakes and indeed how long it will be before these effects manifest on the quality and health of the lakes.

Table 11.5 Goals of strategy for the Rotorua lakes

Protection
• Address the causes of lake water pollution.
• Deal with pollution from septic tanks.
• Determine the extent of pollution from storm-water run-off.
• Define and refine lake water quality standards.
• Examine the status and future of the catchment bank protection scheme
• Address plant and animal pest problems.
• Determine present and future reserve areas.
Use
• Establish an urban development policy.
• Establish a rural development policy.
Enjoyment
• Develop a recreation strategy.
• Monitor and report on recreation activities.
• Define esplanade reserve areas to ensure public access to each lake.
Management goals
• Establish in partnership with Te Arawa a co-management framework that achieves the best integrated management.
• Establish meaningful and binding working relationships with the iwi/hapu and their ancestral lakes.

Source: Hamilton (2003)

Competition

One of the major issues for lakes to contend with is the pressure placed on the resource when urban environments are situated on the lakeshore or in the lakes catchment area. Humans tend to be located disproportionately close to waterways and extensively alter the riparian zones around those waterways (Sala *et al.* 2000). This locational strategy has grown in great part because of the development of a positive aesthetic response to waterfront locations, as well as to the availability of potential recreational activities. These demands have, in turn, led to real-estate development that reinforces such locational drives. Managers must therefore be aware of the needs and requirements of the local community, commercial users, tourists and recreationists, and the 'value' of the lake environment in terms of property prices and development, given the competition that therefore develops for use of such lacustrian space.

Different stakeholders have differing opinions of the aesthetic, social and economic values of a lake and its environs. It is interesting to consider the short-term enjoyment and long-term residential aesthetic appeals of lake environments. In some ways, the way in which the resource is utilised relates strongly to the type of development that would be seen as acceptable and what individual people's perceptions of the state of the environment actually is (Cordella *et al.* 1993). One person may see an environment as pristine and natural, while another may see it as a desolate wasteland (Ryder 1990), owing much to that person's perception or way of measuring health – by the quality of the water they drink or the variety of flora and fauna (Ryder 1990). In addition, conflict and competition between stakeholders can develop because of different understandings of rights to access and use water resources. For example, within the New Zealand framework of property rights with respect to water it is not the property that is owned but the rights to use the property in various ways. In New Zealand. this can be a significant factor in competition between water users because the rights to use water for extractive purposes such as irrigation or energy generation may affect other stakeholders' capacities to engage in non-extractive use such as water-based recreational activities (Harris Consulting The AgriBusiness Group 2003).

Management

Effective management is critical for maintaining the stock of natural resources. However, in the case of New Zealand, the management of lakes is a relatively recent phenomenon as the data collected on lakes continues to grow and develop and better management guidelines and strategies are produced to assist lake managers and users (e.g. Department of

Conservation 2000, 2002). For example, the Waiau Fisheries and Enhancement Trust, a group entrusted with the safeguarding of Lakes Manapouri, Te Anau and Monowai in the South Island of New Zealand, have funded a Ph.D. study to consider the effects of lake-level fluctuations on fish populations to assist in future management decisions (James et al. 2002). The lakes all boast significant trout-fishing opportunities, as well as being important sources of recreational and aesthetic benefits. Indeed, in the New Zealand situation, management attention is arguably focused more on the fishing dimensions of lake use than on any other form of recreational or tourism activity.

Given the lack of appropriate data, there is also a major struggle to identify the appropriate management regimes for certain lakes and to match this to the mix of the range of uses and users of lakes. A recent case is the lake level of Lake Wakatipu, an important lake for recreational activity for the tourist town of Queenstown, in the South Island. A flood mitigation strategy of lowering the lake to its winter mean level has the potential to influence the enjoyment of a local recreational beach (McFall 2003), the fear being that the lowered lake level would 'transform the Frankton beach recreation reserve into a "cesspool of mud, flies and weed"' (McFall 2003: A15). The problems cited include 'the view for tourists arriving in Queenstown would be horrific' and 'the loss of scenic values, quite apart from property values' to the 'natural beauty of the lake and surroundings' (McFall 2003: A15). In addition, the lake at such a low level may also see effluent from the oxidation ponds backwashing into the lake (McFall 2003). Perhaps just as importantly, this example illustrates the interconnectedness of management issues in terms the of technical and social dimensions of foreshore management. Just as significantly, it highlights the issue of the inappropriate location of settlements and infrastructure.

New Zealand's regulatory environment for water resources has developed through reactive need rather than proactive management, often providing a 'tinkering' approach to legislation rather than a coordinated approach to water resource management (Wheen 2002). Traditionally, the development of legislation followed from the adequate use of resources for the needs of settlements, including water supply, road and bridge provision, disposal of sewage and electricity (Wheen 2002). When protection of natural resources was included in legislation, it was only as a matter of human health safety or where economic interests were at stake (Wheen 2002). Indeed, until the 1970s any legislation that focused on recreational and scenic values, economic use rather than direct preservation was favoured (Wheen 2002).

Tourism, leisure and recreation have not heavily influenced policy with regard to soil and water. However, it was in 1981 that an amendment would see the social importance of water and soil conservation given some degree of legal recognition. The Wild and Scenic Rivers amendment (1981) set out 'to recognise and sustain amenity afforded by waters in their natural state' (McColl & Ward 1987: 455). It provided a tool for water levels, flows and quality for waters having 'outstanding wild, scenic, recreational, scientific or other characteristics' (McColl & Ward 1987: 455). This came from growing pressure from environmental and recreational groups towards strengthening environmental management and achieving a greater balance between development and the protection of New Zealand's natural resources (McColl & Ward 1987). As McColl and Ward (1987: 455) state, 'this demand probably stems from growing awareness of the shrinking resource of native forest, wetland, wildlife, unmodified natural water and of the enormous recreational and tourist value of these resources'. If one was taking a more optimistic stance, it may be regarded as an indication that New Zealand society began to view its relationship with nature as being that of harmony rather than domination, with nature being viewed as an intrinsic resource (McColl & Ward 1987).

In the late 1950s, Lakes Manapouri and Te Anau were earmarked as a resource for aluminium smelting in addition to power generation (Peat 1994), although initial discussions of the lakes' generating abilities had begun as early as 1903 (Wheen 2002). The main resource management problem was that the extra generation required through the power station would see the lakes rising up to 30 metres above their normal ranges (Peat 1994).

Two main features of public opinion that affected the case were scenic value and hydroelectricity (Peat 1994). One could not argue the lakes' scenic value and this certainly heightened the impassioned plea by New Zealanders towards lakes that border on national parks and indeed border New Zealand's first World Heritage area (Peat 1994). In addition, the industrialisation of the country saw great interest in the hydroelectriciticpower generation of the southern lakes, which were therefore tied up closely with the country's economic and social prosperity.

While economic and engineering impacts were at the base of the lake-raising issue, what was more important was the social value of the lakes, 'what captured the public's imagination across the country, though, was the prospect that a lake as beautiful as Manapouri could be interfered with, despoiled and debased' (Peat 1994: 3). Nevertheless, talk of conservation and environmental practice were still some years away (Wheen 2002). In 1963, the government decided to 'help itself' and established an enacting piece of legislation to allow it to develop the hydro resource itself, after initial negotiations with a power company in Australia failed (Wheen 2002). Yet by the 1970s, public opinion over Lakes Manapouri and Te Anau solicited a strong commitment by the government to protect the lake resources of the South Island from industrial development.

However, perhaps ironically, new hydroelectric development measures were put in place following the government's decision with respect to Lake Manapouri, which meant the damming of the Clutha, near Clyde in Central Otago, and the creation of a new lake, with substantial recreational and real-estate benefits, which are still being felt today.

The Manapouri campaign cannot be disassociated with the development of environmental law in New Zealand. However, it was also a product of the times, as worldwide interest grew in environmental management (Wheen 2002). Certainly, the 1981 amendment had 'a direct impact on New Zealand environment law' (Wheen 2002: 265).

The more recent Resource Management Act (1991) (RMA), in theory, was to provide for better management of New Zealand's natural resources by removing overlapping jurisdictional regulations and attempting to provide a clearer path for resource management and sustainability. In reality, however, planning issues still remain contested by different interests and individuals who hold different sets of values with respect to natural resource management. Lake management has not been made any easier in New Zealand following the RMA, and the institutional arrangements for lake management remain confused, despite commitment from many for more sustainable forms of development. This situation has only become further complicated by broader contestation over land and foreshore rights in New Zealand between Maori, the Crown and other interested parties. The ownership of rivers, streams, lakes and water is a complex arena of property law.

Under the RMA the retained rights to fresh-water represent those in which various other stakeholders in society have an interest – such as the ecological, fishery and amenity values – and are generally allowed for in the RMA as effects that need to be taken into account in the management of the water resource (Harris Consulting 2003). Nevertheless, some of these effects are regarded as having a higher status than others. The effects of high status are those included in Section 5 (environmental and sustainability issues) and Section 6 (matters of national importance). Other matters in Section 7, such as kaitiakitanga, amenity, intrinsic ecosystem values, environmental quality, recreational and fishing interests and the trout and salmon habitat have lower status under the act as matters to which the consenting authorities should have 'particular regard', rather than matters that must be provided for (Harris Consulting 2003). As Harris Consulting observed in their review of stakeholders' understanding and behaviour with respect to property rights in fresh-water,

> the fact that non extractive stakeholders are not seen as having existing property rights in the water resource with the same level of protection as other existing use rights was a significant bugbear

of stakeholders spoken to. These rights are generally exercised through the planning and consent hearing framework. (2003: 14)

Ownership of lakes may be based on legal title or on common-law rights. The beds of lakes situated completely within boundaries of one parcel of land are vested in the owner of that land unless the lake is specifically excluded from the certificate of title (Hinde et al. 1997). Therefore, the Crown (the New Zealand government) is the owner of many lakes, especially within protected areas such as national parks and nature reserves as well as leased land in the South Island High Country. However, where these conditions are unmet, ownership may be uncertain, and in some case *ad medium filum aquae* rights may apply (Hinde et al. 1997). The beds of some large lakes, such as Lake Taupo, have been vested in the Crown via statute, although it must be noted that the ownership of some lakes remains subject to claims under the Treaty of Waitangi (Statistics New Zealand 2000). For example, ownership of the Lake Taupo bed has been ceded by the crown to the Maori. Other lake resources such as rights to take fish and gamebirds is restricted under the Conservation Act and the Wildlife Act (1953) respectively. In New Zealand, unlike many other countries, commercial taking of freshwater fish and game is extremely limited, although fishing and hunting are major economic uses of lakes and wetland areas vis licensing, servicing and guiding activities. The economic, social and environmental dimensions of tourism and recreation have only further complicated the legal aspects of ownership and responsibility, yet remain relatively intangible dimensions in the planning process.

Undoubtedly, sustainable development has become an important concept or ideal for managing the environment for the New Zealand government, stakeholders and the tourism industry (Dymond 1997; Hall & Kearsley 2001). Many governmental policies have altered their approach to management by applying the concept of sustainability in order 'to reconcile the competing claims of economic development and conservation by emphasising the need to bequeath to future generations an undiminished environmental resource' (Hughes 2002: 461). The RMA has been the main instrument in developing sustainability, while national tourism strategies have also started to recognise the concept of sustainable development (Hall & Kearsley 2001). Increased knowledge of environmental systems and ecosystems has also encouraged policy settings to shift attention from managing a single element to managing all elements of an ecosystem or watershed. Unfortunately, in the case of lake management in New Zealand, while ecosystem approaches to sustainable lake management are to be applauded, there is still much to learn and twice as much to implement to achieve the goals of sustainable ecosystem management. As the New Zealand Minister for the

Environment, Marion Hobbs, stated, 'until recently we have taken our abundance of freshwater in New Zealand for granted' (in Water Programme for Action Inter-departmental Working Group 2004: 2). Critical to this is the development of more cooperative and adaptive management procedures that are able to bring together the key stakeholders in the complex legislative and policy environment that surrounds many of the country's lakes and freshwater systems, as even the best understanding of lacustrine systems will be without value unless such knowledge can be implemented. For example, with respect to the Strategy for the Lakes of the Rotorus District discussed above, implementation requires a multiple agency approach with overall coordination and direction for the strategy being vested in the Lakes Strategy Joint Committee, comprising the Chair of the Te Arawa Maori Trust Board, the Chair of Environment Bay of Plenty and the Mayor of Rotorua District Council, plus a further representative from each of the three organisations (Hamilton 2003).

The Rotorua and Taupo initiatives are a good example of increasing cooperative arrangements by various stakeholders in complex water-resource management environments. However, such developments also occur in other sectors and locations with respect to the goals of maintaining or improving frehwater quality, especially with respect to the impact of farming and the effects of sewage systems. For example, beyond the specific regulatory approaches that are available under the RMA, remedial strategies in the dairy sector for mitigating adverse environmental effects tend to focus on voluntary codes of conduct and good practice, including a May 2003 Dairying and Clean Streams Accord that has been signed by Fonterra Co-operative Group (New Zealand's largest dairying group), Local Government New Zealand, Ministry for the Environment, and Ministry of Agriculture and Forestry (PCE 2004). The purpose of the accord is to prove a 'statement of intent and framework for actions to promote sustainable dairy farming in New Zealand. It focuses on reducing the impacts of dairying on the quality of New Zealand streams, rivers, lakes, groundwater and wetlands' (PCE 2004: 137) with particular benefits not only for stock water, but also for fish and swimming. Under the accord, dairy cattle are to be excluded from 50% of streams, rivers and lakes by 2007 and 90% by 2012. According to the PCE (2004: 137), by August 2004, 'nine of the 13 regional councils have developed clean water action plans under the Accord. The plans cover 84% of Fonterra's farmer suppliers.'

Despite the value of the various collaborative arrangements that have been entered into, it needs to be recognised that they are primarily reactive in nature and that while they have potentially significant benefits for lake-based tourism and recreation, tourism is an incidental beneficiary rather than an integral component of lacustrine and catchment resource management. Nevertheless, a recent freshwater plan of action developed by the national government has at least started to make some segments of government, at all levels, and possibly the general public, aware of some of the issues associated with the integrated management of freshwater resources.

Water Programme of Action

In 2003, the New Zealand Ministry for the Environment and the Ministry of Agriculture and Forestry jointly established a Water Programme of Action to examine the pressures and competing interests for New Zealand's freshwater resources. The programme is part of the national government's broader Sustainable Development Programme of Action. The Water Programme of Action involved many government departments, including the Ministry of Tourism, as well as involving representatives of regional councils and other local authorities, a Mäori Reference Group and a Stakeholder Reference Group. The vision for the programme is that 'Freshwater is managed wisely to provide for present social, cultural, environmental and economic wellbeing of New Zealand' (OME 2004), with key challenges being recognised as:

- not all expectations and needs for freshwater are currently being met and demands are growing;
- water quality is declining in many areas and is unacceptable in some catchments;
- given the range of people's interests in water it is difficult to establish priorities for action. (OME 2004; MFE 2004c)

The objectives and principles for the sustainable development of fresh water under the Programme of Action is contained in Table 11.6. Work undertaken as part of the programme has identified issues surrounding water allocation and use, water quality and the identification of potential water bodies of national importance as being significant elements of the programme, as well as a preferred package of actions that aim to enhance freshwater resource management through confirming the role of local government in freshwater management and decision-making, but with greater direction and support from central government (OME 2004). The preferred package of actions is listed in Table 11.7. The preferred package of actions was included in a discussion paper (MFE 2004c) that was made available for public consultation and discussion in December 2004. Submissions closed on 18 March 2005 with the receipt of approximately 300 submissions. At the time of writing, submissions were being collated with the government moving towards the next stage of the programme (updates were scheduled to be provided at www.mfe.govt.nz/issues/water/prog-action/index.html). However, with a national election due

Table 11.6 Objectives and principles for New Zealand's water Programme of Action

Objectives
• Protect public health
• Facilitate economic growth and innovation
• Facilitate public use, access, and enjoyment
• Enhance environmental protection
• Manage freshwater in the context of Maori cultural values and the treaty relationship between Crown and Maori
Principles to achieve objectives
• Decision-making should be transparent, participatory and timely
• Manage within the constraints of uncertainty and cost
• Respect existing rights, interests and values and future options
• Maintain environmental bottom lines and avoid, remedy or mitigate adverse environmental effects
• Decision-making should occur at the appropriate level and balance local and national interests
• Decision-making is underpinned by adequate information
• Water is made available over time for its highest value use

Source: Office of the Minister for the Environment (2004)

before the end of 2005, the next and arguably most controversial stage with respect to implementation of the programme and the development of new regulatory mechanisms, is likely not to commence until a new government is formed. Moreover, while tourism is likely to be a key beneficiary of the proposed programme of action, it is noteworthy that tourism was barely mentioned in the discussion document (MFE 2004c) and that understanding of how to integrate tourism and recreational concerns into freshwater management, except with reference to water quality, continues to be poor.

Conclusion

Lakes and their surrounding environments, or the lacustrine environment, are important resources for tourism, leisure and recreation activities in New Zealand, as well as supporting an indefinable list of personal benefits and values. Issues involving lakes and lake management strategies that attempt to mitigate the negative aspects of lake usage

Table 11.7 Summary of preferred package of actions with respect to fresh water in New Zealand

i.	*Develop National Policy Statements* that specify national priorities for freshwater, stipulate requirements of regional plans, and require catchment-based targets for water quality to be set.
ii.	*Develop National Environmental Standards* to specify methods or procedures for setting environmental bottom lines and allocation limits, and to address the management of diffuse discharges.
iii.	*Address nationally important values* by identifying water bodies with such values and making this information widely available, and by prioritising for action water bodies with nationally important values that are under threat.
iv.	*Increase central government participation in regional planning.* This could include providing information and guidance, and lodging submissions by either individual departments or through the whole of government process.
v.	*Increase central government support for local government* by capacity building or disseminating good practice for: strategic planning processes for freshwater; setting environmental bottom lines and allocation limits; engaging effectively with Maori; managing clawback and transfer of water permits; and enhancing efficiency of water use.
vi.	*Develop special mechanisms for regional councils,* including powers to progressively constrain (clawback) existing consents to take water or discharge contaminants where water is over-allocated or water quality is declining.
vii.	*Enhance the transfer of allocated water between users.* Mechanisms could include developing a pilot registry system which regional councils could choose to use to record water transfers, and working with local government to encourage greater consideration of transfer of water.
viii.	*Develop market mechanisms to manage discharges,* including mechanisms to trade permissions to discharge particular contaminants.
ix.	*Set requirements for regional freshwater plans to address key issues and challenges* in areas where water resources are under pressure, and promote the implementation of regional plans to also be reflected in long-term council community plans.
x.	*Enhance Maori participation* where they have an interest, by strengthening involvement of Maori in national and regional

Table 11.7 continued

	strategic planning and providing central government support for better engagement.
xi.	*Enable regional councils to allocate water* to priority uses by allowing applications for resource consents to be heard on a comparative basis, allowing regional councils to identify priority uses for water and develop criteria to guide decision making on the allocation of abstracted water, and allowing regional councils to use market tools (e.g. tendering, auctioning) as one of the ways of allocating water within the comparative framework.
xii.	*Raise awareness of freshwater problems and pressures,* and promote solutions for managing the impacts of land use on water quality, over-allocation and inefficient water use. This could include communication and education programmes, and development of voluntary agreements.
xiii.	*Work with local government, scientists and key stakeholders* on pilot projects to demonstrate and test new water management initiatives.

Source: After Office of the Minister for the Environment (2004); Ministry for the Environment (2004c)

while maximising enjoyment and benefits for humans are widespread. In addition, the 'legislative, social and economic factors interact to change both human attitudes and behaviours towards lakes' (Walsh *et al.* 2003: 212).

In the case of New Zealand, research thus far has indicated a relatively poor reference base from which to develop sustainable lake management practices, although recent government initiatives with respect to freshwater resources at least start raising stakeholder awareness about many issues related to the environmental qualities of lakes for recreation and tourism. As the Ministry of Tourism (2004: 2) has stated with respect to the role of iconic lake and river images in the promotion of New Zealand as '100% Pure', 'It is vital that New Zealand delivers on the "promises" made in these campaigns and the sustainable management of our freshwater bodies is part and parcel of this.' However, the base-line level of knowledge is poor, while the institutional arrangements for lake management are confused. Nevertheless, use of lakes for a variety of purposes is growing, with potential conflicts between users on the rise. It is indeed ironic that for a resource that is so integral a component of New Zealand's clean and green image, there is relatively so little understanding of what the long-term implications of increased tourist and recreational use might actually be.

References

Cordella, P., Salmaso, N. and Cavolo, F. (1993) Findings and suggestions from research in Lake Garda. In G. Giussani and C. Callieri (eds) *Strategies for Lake Ecosystems and Beyond 2000, 5th International Conference on the Conservation and Management of Lakes* (pp. 471–4). Lake Biwa, Japan: International Lake Environment Committee Foundation.

Davies-Colley, R., Cameron, K., Francis, G., Bidwell, V., Ball, A. and Pang, L. (2003) *Effects of Rural Land-use on Water Quality*. Hamilton: NIWA.

Davis, S.F. (1987) *Wetlands of National Importance to Fisheries*. New Zealand Freshwater Fisheries Report No. 90. Christchurch: Freshwater Fisheries Centre, MAFFish.

Department of Conservation (2000) *Lake Wairarapa Wetlands Action Plan 2000–2010*. Wellington: Department of Conservation.

Department of Conservation (2002) *Pencarrow Lakes: Conservation Values and Management*. Wellington: Department of Conservation.

Dymond, S.J. (1997) Indicators of sustainable tourism in New Zealand: A local government perspective. *Journal of Sustainable Tourism* 5 (4), 279–93.

Hall, C.M. and Kearsley, G. (2001) *Tourism in New Zealand: An Introduction*. Melbourne: Oxford University Press.

Hamilton, B. (2003) *A Review of Short-term Management Options for Lakes Rotorua and Rotoiti*. Report prepared for the Ministry for the Environment. Wellington: Ministry for the Environment.

Harris Consulting The AgriBusiness Group (2003) *Property Rights in Water: A Review of Stakeholder's Understanding and Behaviour*. Report prepared for MAF Policy and Ministry for the Environment. Wellington: Ministry for the Environment.

Hinde, G.W., McMorland, D.W., Campbell, N.R. et al. (1997) *Butterworths Land Law in New Zealand*. Wellington: Butterworths.

Hobbs, M. (2004) Funding to Imporve Lake Rotoiti Water Quality. Press release, 26 June.

House, M.A. (1996) Public perception and water quality management. *Water Science and Technology* 34 (12), 25–32.

Hughes, G. (2002) Environmental indicators. *Annals of Tourism Research* 29 (2), 457–77.

James, M., Mark, A. and Single, M. (2002) *Lake Managers' Handbook. Lake Level Management*. Wellington: Ministry for the Environment.

Larned, S., Scarsbrook, M., Snelder, T. and Norton, N. (2005) *Nationwide and Regional State and Trends in River Water Quality 1996–2002*. Prepared for the Ministry for the Environment by National Institute of Water & Atmospheric Research Ltd. Wellington: Ministry for the Environment.

McColl, R.H.S. and Ward, J.C. (1987) The use of water resources. In A.B. Viner (ed.) *Inland Waters of New Zealand* (pp. 441–50). DSIR Bulletin 241, Wellington.

McFall, L. (2003) Lower levels in Wakatipu could lead to mudflats. *Otago Daily Times*, 22 February, A15.

Ministry for the Environment (MFE) (2004a) *Lake Taupo: The Ministry for the Environment's Perspective*. Wellington: Ministry for the Environment.

Ministry for the Environment (MFE) (2004b) *Water Bodies of National Importance: Potential Water Bodies of National Importance for Recreation Value*. Wellington: Ministry for the Environment.

Ministry for the Environment (MFE) (2004c) *Freshwater for a Sustainable Future: Issues and Options: A Public Discussion Paper on the Management of New Zealand's Freshwater Resources*. Prepared for the Minister for the Environment by the Water Programme of Action Inter-departmental Working Group. Wellington: Ministry for the Environment.

Ministry for the Environment and Ministry of Health (2002) *Freshwater Microbiology Research Programme Report: Pathogen Occurrence and Human Health Risk Assessment Analysis*. Wellington: Ministry for the Environment.

Ministry of Agriculture and Forestry (MAF) (1993) *Sustainable Agriculture: MAF Policy Position Paper 1*. Wellington: Ministry of Agriculture and Forestry.

Ministry of Tourism (2004) *Subject: Waters of National Importance for Tourism Report*. Wellington: Ministry of Tourism.

Mosley, P. and Jowett, I. (1999) River morphology and management in New Zealand. *Progress in Physical Geography* 23 (4), 541–65.

Office of the Minister for the Environment (OME) (2004) *Cabinet Paper: Water Programme of Action – Consultation on Policy Direction, POL (04) 320*. Wellington: Office of the Minister for the Environment, Cabinet Policy Committee.

Parliamentary Commissioner for the Environment Intensive Farming (PCE) (2004) *Sustainability and New Zealand's Environment*. Wellington: Parliamentary Commissioner for the Environment.

Peat, N. (1994) *Manapouri Saved: New Zealand's First Great Conservation Success Story*. Dunedin: Longacre Press.

Quinn, J.M. and Stroud, M.J. (2002) Water quality and sediment and nutrient export from New Zealand hill-catchments of contrasting land-use. *New Zealand Journal of Marine and Freshwater Research* 36, 409–29.

Rosen, M.R. (2001) Hydrochemistry of New Zealand's aquifers. In M.R. Rosen and P.A. White (eds) *Groundwaters of New Zealand* (pp. 77–110). Wellington: New Zealand Hydrological Society Inc.

Ryan, P.A. (1991) Environmental effects of sediment on New Zealand streams: A review. *New Zealand Journal of Marine and Freshwater Research* 25, 207–21.

Ryder, R.A. (1990) Commentary. Ecosystem health, a human perception: Definition, detection and the dichotomous key. *Journal of Great Lakes Research* 16, 619–24.

Sala, O.E., Chapin, F.S., Armesto, J.J., Berlow, E., Bloomfield, J., Dirzo, R., Huber-Sanwald, E., Huenneke, L.F., Jackson, R.B., Kinzig, A., Leemans, R., Lodge, D.M., Mooney, H.A., Oesterheld, M., Poff, N.L., Sykes, M.T., Walker, B.H., Walker, M. and Wall, D.H. (2000) Global biodiversity scenarios for the year 2010. *BioScience* 287, 1770–74.

Scott, W.G., Scott, H.M., Lake, R.J. and Baker, M.G. (2000) Economic cost to New Zealand of foodborne infectious disease. *New Zealand Medical Journal* 113, 281–4.

Smith, D.G., McBride, G.B., Bryers, G.G., Wisse, J. and Mink, D.F.J. (1996) Trends in New Zealand's national river water quality network. *New Zealand Journal of Marine and Freshwater Research* 30, 485–500.

Statistics New Zealand (2000) *New Zealand Official Yearbook 2000*, 101st edn. Auckland: David Bateman.

Tourism New Zealand (2003a) What visitors want: Product marketing presentation April 2003. Available online: www.tourisminfo.govt.nz/cir_pub/index.cfm?fuseaction=77. Accessed 14 May 2003.

Tourism New Zealand (2003b) Long hot summer for tourism. Available online: www.tourisminfo.govt.nz/cir_news/index.cfm?fuseaction=newscentre&subaction=news&article_id=496. Accessed 20 February 2003.

Tourism Research Council of New Zealand (2003). Domestic travel survey. Available at www.trcnz.govt.nz. Accessed 31 May 2003.

Unwin, M. and Brown, S. (1998) *The Geography of Freshwater Angling in New Zealand: A Summary of Results from the 1994/96 National Angling Survey.* Christchurch: NIWA for Fish and Game New Zealand.

Unwin, M. and Image, K. (2003) *Angler Usage of Lake and River Fisheries Managed by Fish and Game New Zealand: Results from the 2001/02 National Angling Survey.* NIWA client report CHC2003–114 December. Christchurch: National Institute of Water and Atmospheric Research Ltd.

Vant, W.N. (2001) New challenges for the management of plant nutrients and pathogens in the Waikato River, New Zealand. *Water Science and Technology* 43 (5), 137–44.

Walsh, S.E., Soranno, P.A. and Rutledge, D.T. (2003) Lakes, wetlands and streams as predictors of land-use/cover distribution. *Environmental Management* 31, 198–214.

Water Programme for Action Inter-departmental Working Group (2004) *Freshwater for a Sustainable Future: Issues and Options. A Public Discussion Paper on the Management of New Zealand's Freshwater Resources.* Prepared for the Minister for the Environment. Wellington: Ministry for the Environment.

Wheen, N.R. (2002) A history of New Zealand environmental law. In E. Pawson and T. Brooking (eds) *Environmental Histories of New Zealand* (pp. 261–74). Melbourne: Oxford University Press.

Wilcock, R.J., Nagels, J.W., Rodda, H.J.E., O'Connor, M.B. *et al.* (1999) Water quality in a lowland stream in a New Zealand dairy farming catchment. *New Zealand Journal of Marine and Freshwater Research* 33, 683–96.

Chapter 12
Local Considerations in Marketing and Developing Lake-destination Areas

DANIEL L. ERKKILÄ

Introduction

It is not unusual for tourism development to happen quite simply. It may only be the coming-together of an idea (product or service) by someone with the know-how and resources to create the business and be successful. That may be appropriate where there are few other players beside the entrepreneur, his or her banker and the new customer. While this situation could exist in any destination, research is suggesting that lake-destination areas and their product development should be different in critical ways.

Certainly, not all major lake-tourism destination areas in the world are the same. It is logical to assume that different locales will be at different stages in the product life-cycle, have different markets and different opportunities. Common themes may be emerging, however, because of the common draw – lakes and water. Further, in the case of lake-destination tourism, other influential groups frequently come into play, suggesting that successful development and marketing of lake-destination areas may be more complicated, requiring careful attention to all stakeholders' needs and obtaining adequate information to make sound decisions.

The primary stakeholders in lake-tourism business development have always been the customer and business providing the touristic good or service. Understanding the customer's motivations for travel, needs and travel preferences have been central to the development of new products and connecting those products to the customer through effective marketing. Lake-destination image and branding are ingredients of the marketing campaign trade. It is often assumed that if the customers are coming, the needs of the business owner are met. Because lakes are frequently found away from large population centres, however, value-added development is often motivated by the desire to attract new visitors to lake regions and to retain them for extended periods to boost expenditures. This also increases host–guest interaction. Secondary stakeholders, then, are host-community residents and, in particular, owners of shoreline property around the marketed lakes. Paying attention to their needs is important as well.

Background: Minnesota as a Lake-destination Example

Upper Midwest US states such as Minnesota are excellent examples of lake-destination regions that can serve to illustrate this discussion. What follows will highlight what the Minnesota research and lake-destination development experience has told us relative to key lake issues, particularly within the framework of stakeholder interests. Important background may first prove helpful to understand the Minnesota experience.

Long branded the 'Land of 10,000 Lakes', Minnesota actually has nearly 12,000 lakes that are 10 acres (4.1 ha) or larger. Most lakes are natural, the result of repeated ancient glacial movement and deposits, and are in the northern half of the state. They are the draw for the major urban settlement of the St Paul–Minneapolis area and urban centres in neighbouring Midwest states. Total inland surface water area is nearly 4900 square miles (12,500 km^2). The largest inland lake is nearly 290,000 acres (117,000 ha) and the state shares over 150 miles (241 km) of Lake Superior shoreline – the largest, coldest and deepest of the Great Lakes and the world's largest freshwater lake in surface area. The first mention of Minnesota's lakes was originally in a newspaper. The first association between Minnesota and lakes in branding terms probably occurred originally in 1858 when a newspaper in what is now Minneapolis called Minnesota the 'Land of Lakes'. When '10,000' was first added is less clear, but it has been documented as far as 50 years ago.

While it is not possible to differentiate between lake tourism and non-lake tourism, the scale of Minnesota tourism in 2001 included 19.2 million pleasure travellers, 3 million business travellers and 0.7 million international travellers (Minnesota Office of Tourism 2002). This economic activity generated US$9.8 billion in gross receipts, 135,000 jobs and US$1.1 billion in state and local tax revenue. Minnesota residents account for 52% of all non-business travellers, with the neighbouring Midwestern states of Wisconsin, Iowa, North Dakota and Illinois leading as the top states of non-resident travellers. Travellers to the northern half of the state account for 36% of total number of visitors to the state.

The Tourist as Stakeholder – Lake-destination Travel Motives and Needs

The explanation of tourist motivations for travel behaviour was simply framed by Dann (1977) using a push–pull framework. Push factors are those that initially motivate us all to take vacations and travel outside our normal environment and pull factors are those forces that draw us to a particular destination (over another) after the decision to travel has been made. Klenosky (2002) summarised past empirical research on push factors as generally relating to the needs and wants of the traveler, including seeking escape, rest and relaxation, adventure, prestige, health and fitness and social interaction. His review of past research on pull factors generally show them as describing the features, attractions or attributes of destinations specifically, including the draw of sunny beaches, good restaurants and inexpensive travel opportunities.

Getting the Traveller to the Region: Trip Motives and Lake Attraction

Research in the lake regions of Wisconsin and Minnesota have revealed similar results regarding the push factors that influence travellers. Norman *et al.* (1995) used a postal survey of 1036 households in four major market areas to profile Wisconsin travellers to a north-central Wisconsin lakes area. Overall, the top reasons cited for taking holiday trips to the Wisconsin lakes areas (Likert scale: 1 = not at all important; 4 = very important) were:

- getting away from demands of home (mean 3.5);
- travelling to places where I feel safe and secure (3.5);
- being together as a family (3.4);
- seeing and experiencing a new destination (3.4);
- visiting a place that takes care of environment (3.3).

These trip reasons were grouped for underlying motives – i.e. push factors – with results shown in Table 12.1. The study also identified the top important pull factors, or importance of facilities, features and services, in selecting a holiday destination in the Wisconsin lakes region, including: 'Outstanding scenery' (mean = 3.6); 'High standards of cleanliness' (3.6); 'The appearance of the area' (3.6); 'Lakes and rivers' (3.5); and 'Friendly local residents' (3.5).

Recent research regarding Minnesota lake-destination areas has also shown 'escape' to be a premier motivation for holiday-travel decisions and the top two holiday-travel destination attributes, or pull factors, for in-state survey respondents were shown to be 'Being near water' (38%)

and 'Relaxation' (35%). Out-of-state respondents rated those features equally (35%). Value-added development opportunities for communities and businesses were suggested by the fact that both in- and out-of-state respondents said 'Good food/restaurants' were the third most important feature for a holiday destination (Colle & McVoy 2003).

Table 12.1 The underlying motives for taking a holiday trip

Underlying motives (and examples of stated reasons)	Mean[a]
Escape • Getting away from the demands of home/job. • Being together as family. • Visiting a place that takes care of its environment.	3.24
Education • Travelling through places that are important in history. • Experiencing a new and different lifestyle. • Trying new food. • Seeing and experiencing a new destination.	3.00
Family • Visiting friends and relatives. • Visiting places my family came from.	2.62
Action • Finding thrills and excitement. • Being daring and adventuresome. • Roughing it. • For the adventure. • To improve my skills in an activity.	
Relaxation • Doing nothing at all. • Indulging in luxury.	2.44
Ego • Going places my friends haven't been. • Talking about the trip after I return home.	2.38

[a] Based on a 4-point Likert scale of 1 = not at all important; 2 = not very important, 3 = somewhat important; 4 = very important.
Source: Norman *et al.* 1995

Image and Brand - Tying it Together

Successful lake-destination area marketing is tied to a strong destination image. As a region, for example, years of consistent work on lakes and water-image formation appears to have succeeded in the minds of potential Minnesota travellers. As shown in Table 12.2, research indicates that the state's water-based attributes are 'top of mind', especially among state residents, and are also very high among non-residents. Similarly, these attributes carry over as key equities in making Minnesota unique as a holiday destination for both in-state residents and those travelling to Minnesota from other regions, as shown in Table 12.3 (Colle & McVoy 2003).

Related research suggests why water is so important for holidays: it is intimately linked to holidaymakers' need for relaxation (Colle &

Table 12.2 'Top of mind' associations with Minnesota

In-state residents	Out-of-state residents
Lakes/rivers/water: 87%	Lakes/rivers/water: 55%
Natural beauty/forests/wilderness: 45%	Winter/cold/snow: 50%
Winter/cold/snow: 35%	Twin cties: 17%
Outdoor activities (e.g. fishing, hiking): 32%	Sports (pro-basketball, etc.): 16%
Twin cities: 20%	Forest/trees/woods: 15%

Source: Colle & McVoy (2003)

Table 12.3 Minnesota's unique equities as a holiday destination

In-state residents	Out-of-state residents
Lakes/rivers/water: 48%	Lakes/rivers/water: 40%
Has a lot of variety: 37%	Natural beauty/outdoors/activities: 36%
Natural beauty/outdoors/activities: 34%	No crowds/can find privacy/getaway: 15%
No crowds/can find privacy/get away: 6%	Seasons/weather: 9%

Source: Colle & McVoy (2003)

McVoy 2003). When asked, respondents described how being near water on holiday made them feel:

- relaxed;
- it's soothing and calming;
- it lulls me into a relaxed frame of mind;
- peaceful and relaxed – love to listen to the waves;
- it's fun for activities like canoeing, fishing and water-skiing;
- I feel closer to nature.

Collectively, this research was tied together in Minnesota for an aligned marketing campaign that linked the predominant motivation for choosing a holiday – 'escape' – with, among others, the lake-destination pull of water and the strong emotions tied to being around water. Armed with a strong water image in travellers' minds, the state marketing agency focused on the emotional benefit of travel and elevated a key rational equity – water – to become a broader, more encompassing vision for the brand. The new brand hierarchy became (in declining order):

- *Promise*
  Exploring Minnesota replenishes what the rest of the world takes away.

- *Personality*
  Genuine, come as you are, captivating, supportive

- *Emotional Benefits*
  A deeper level of contentment.
  The vitality that springs from refreshing and reconnecting with yourself.
  The simple joy of being with family and friends.

- *Differentiating Rational Benefits*
  Unspoiled settings, never far from the serenity of water.
  Intense seasons that reveal nature's most magical moments.
  Discovery of best-of class-experiences.

For a marketing campaign, this combination of focusing on the customer over the place and emotional response over rational choice placed Minnesota's marketing opportunities in a unique position, relative to other neighbouring states (Figure 12.1). As a new campaign, its success has not yet been determined; but, because it was in development prior to the 9/11 terrorist attacks in New York, it appears to be ready-made for the sentiments of post-9/11 travellers.

Marketing Lake-destination Areas

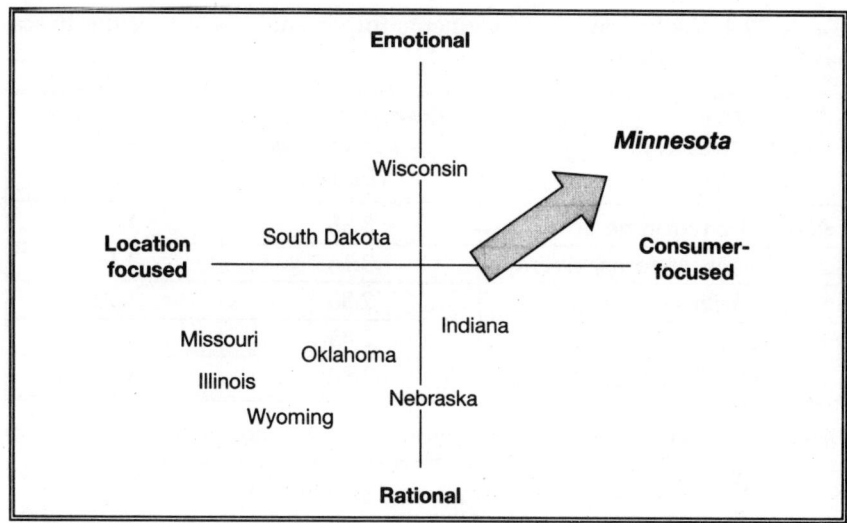

Figure 12.1 Minnesota's marketing opportunity map

Localized Traveller Expectations: The Importance of Customer Profile Information

At the local lake-destination level, recent research corroborates the macro-level research in Minnesota. A year-long intercept survey of more than 2100 travellers to lake-destination area of north-east Minnesota (University of Minnesota Tourism Centre 2002) focused on a regional destination centre known for many resource-based activities. These include leisure and business travel, with natural resource endowments involving 1.3 million acres (540,000 ha) of forestland, the Mississippi River and more than 1000 lakes over 10 acres in size (> 4 ha). Here, once the state and regional marketing initiatives get the visitor on-site, measuring and managing tourist satisfaction and needs were found to be very important for sustained local success.

The north-east Minnesota research showed that visitors came to the lake destination area for a wide variety of reasons, including outdoor recreation (20%), destination business (19%) and staying with friends or relatives (14%). Pull factors were identified and visitors were asked how well the area provided identified attraction attributes (Table 12.4). The top elements visitors identified as *most important* in selecting the area were: 'The natural environment' and 'Area lakes, streams and rivers'. Visitors were *most satisfied* with 'The natural environment' and 'Area lakes, streams and rivers'. These findings suggest that in this lake destination area, there is a very good match between elements that visitors

Table 12.4 Visitor ratings of elements important in selecting the Itasca lakes area as a destination

Itasca area (n = 1603–2001)	Mean importance to destination choice	Mean rating of experience satisfaction
Natural environment	3.36	3.53
Area lakes, streams, rivers	3.16	3.46
Scenic drives	2.86	3.30
Fishing/hunting	2.85	3.23
Boating/water recreation	2.80	3.39

Based on a 4-point Likert scale of importance: 1 = not important, 4 = very important. Satisfaction Scale: 1 = not satisfied, 4 = very satisfied
Source: University of Minnesota Tourism Center (2002)

consider important when *choosing* the area and their *satisfaction* with those elements when they experience them.

Measurement of other reasons for people travelling to an area also revealed visitor needs that identified potential new development or marketing opportunities. For example, visitors to northern Minnesota lake areas stated that the single most important source of information that contributed to their awareness of the destination area was obtained by previous visits, followed by recommendations from friends and relatives (University of Minnesota Tourism Centre 2001b, 2002). Previous visits and recommendations from friends and relatives were also the two most important information sources for planning holiday trips. This suggests that customer satisfaction with first-time visits is critical for repeat visitation and, in this case, local resident awareness for travel options within an area is important in conveying recommendations to friends and family who might be interested in visiting a lake destination.

This same customer profile research in the lake destination areas of northern Minnesota showed that visitors generally spend an average of 33% of total expenditures on lodging, 19% on dining and 19% on retail shopping. With accommodation being the largest component of daily expenditures, research revealed that visitors rated 'Good value for the price' as the highest amenity in importance for lodging site selection, but 'Economy priced,' 'Nature setting' and 'On a lake or river' followed in relative importance. At the local level, it was discovered that visitors were generally focused on value and proximity to water, and were interested in upscale offerings such as cable TV, indoor pools and proximity to area attractions. Based on this customer profile research, the new information will guide investors interested in future development opportunities towards upgrading area facilities to increase visitor expenditure.

The Tourist Enterprise as Stakeholder: The Provider's Needs

The front-line providers of goods or services for visitors to a lake-destination area visitor are typically the small businesses in the region. They generate the value-added economic development that provides jobs and income to the area. Business success stems from applying sound business principles and, of course, a keen sense of customer preferences.

One way this is being done in a formal, organised fashion is through business retention and expansion programmes as used in Minnesota and elsewhere. These programmes focus on the varied needs of private enterprises working in communities, (see *Business Retention & Expansion International* website and resources at www.brei.org). Specific objectives of these programmes, as applied to a lake destination area, would include:

- illustrating the economic importance of the travel and tourism industry to the area;
- demonstrating the area's appreciation of area tourism businesses for their contributions to the local economy;
- developing action plans to remove obstacles for continued and balanced growth of the lake destination area's travel and tourism industry. (Erkkilä 1995)

One business programme (Erkkilä and Morse 1995) in a north-east Minnesota lake area involved on-site visitation and survey of 73 predominantly lake-based lodging operations and businesses offering attractions to the area. Assessment of the economic health of firms, labour market issues, employee productivity and training needs, constraints on expansion or renovation and information needs resulted in key findings, with examples including:

- *Employee training* As a service industry increasingly dependent on service quality to ensure repeat business, surveyed firms rated their employees in numerous categories (e.g. customer service, knowledge of tourists' needs) and saw the benefit of developing employee training plans.
- *Constraints on renovation or expansion* Firms identified constraints to growth (e.g. government land-use planning issues) to open a dialogue to remedy the situation with area leaders.

- *Business information needs* Firms were eager for information/educational resources (e.g. marketing planning information), setting an agenda to work with neighbouring businesses and university resources for workshops and other assistance.
- *Community services* Firms recognised the need for strong regional infrastructure support (e.g. ambulance, police, fire), again setting the stage for a dialogue with regional officials on particular issues.
- *Quality of related businesses* As a group, tourism firms rated their peers to ascertain whether they were considered as strong or weak in supporting the tourist. Those who were considered weak links in the distribution system were identified for improvement.
- *Promotion and advertising* Firms identified the needs for expanding promotional efforts for travel and tourism, and their preferences for various approaches to raise the finance to do so.

Importantly, this tourism business research has shown that lake area firms in Minnesota appear to be in agreement with their customers regarding the of pull factors. As tourists stated they were drawn by the 'Natural environment and area lakes' and 'Streams and rivers.' Tourism businesses independently stated that the main attractions to visitors were 'Access to lakes and rivers' (93%) and an 'Unpolluted natural environment' (90%). Agreement between providers and consumers is important for customer satisfaction and repeat business.

Experience with organised, programmatic attention to commercial needs using an approach such as tourism business retention and expansion programmes has uncovered positive outcomes. Apart from identifying the expected needs of firms to have sound customer profile research and tourism marketing data, organisations in lake areas that spend time on this work can generate strategic plans for providing resources to tourism firms to: strengthen business knowledge; build stronger relationships with related businesses for new growth; better connect with public/government partners to remove barriers to economic development, and grow employment opportunities over the long term.

The Lake Destination Community as Stakeholder: Playing Host

In situations where lake destinations are neighbouring or in close proximity to significant inhabited areas, attention needs to be paid to the host community that frequently provides guest services or goods to travellers as they drive to, or return from, lake-destination activities. As service centres and locations for new value-added opportunities, the valuable guest–host relationship should be remembered for the relationship to stay strong and positive.

Community Tourism Development

The context for host-community consideration in tourism development is the need for the industry to remain in agreement with residents – i.e. when development projects are being planned, the values, needs and goals of the community should be understood and taken into consideration. Research in this area includes the study of residents' quality of life and their attitudes towards tourists (Jurowski 1994; Gursoy et al. 2002). Resources are available to assist communities in identifying the key values for tourism development (e.g. University of Minnesota Tourism Centre 2001a), such as:

- *Commitment to assessing the true benefits and costs* Consideration of the social, economic and environmental benefits and costs.
- *Diversification* Tourism as a diversification tool, not a substitute for other local industry.
- *Quality* Sustenance of quality products and services to avoid the loss of visitors.
- *Shared benefits and control* Business and community residents sharing the benefits and control of future tourism development.
- *Authenticity* Staying true to the social fabric, history and culture of the community – the attributes that make the community unique.

Resident Perceptions and Needs

For lake-destination areas, the sustainability of resources and tourism products/services is critical for achieving and maintaining a balance between stakeholder needs. The research in north-east Minnesota also involved a random sample postal survey sent to 1370 residents out of a population of 15,000 resident homes to gain an understanding of the perspective of the resident population towards tourism (University of Minnesota Tourism Centre 2002).

Residents indicated that the main the number one attraction that most enhances the lives of residents and tourists was 'Outdoor recreation' (58% of all respondents). Other findings that will inform future developers in lake area destinations included:

- Top concerns to tourism growth were higher taxes (21%), overuse of outdoor areas (13%) and increased pollution (10%).
- Top benefits to tourism growth were more healthy economies (32%), employment opportunities (26%) and a greater variety of goods and services to residents (14%).
- Residents agreed that tourism increased the quality of life in the area and helped to balance the economy, but future growth should be controlled.

This study also involved focus group work to define further the community's values within which tourism development occurs. Results revealed that residents in lake-destination areas consistently rated 'Respect for the environment', 'Authentic history and heritage' and a 'Customer service orientation' as positive community characteristics that they valued highly. Residents said these values were important (host-destination) values that they expected to find while travelling away from home themselves, as well as what they felt should be in place in their own community as it prepares to host new visitors.

As development occurs, it is also important to be aware that not all lake-area residents are year-round ones and that seasonal resident perceptions of tourism growth may differ from those of permanent residents. A comparison among respondents that lived in the north-east Minnesota study area year-round versus those who had permanent residences elsewhere but came seasonally, showed differences in attitudes. Here, 87% indicated year-round status, while 13% were seasonal. (As a group, the seasonal residents tended to be slightly older than year-round residents and more affluent.) Seasonal residents, compared to year-round residents, tended to:

- be less supportive of the same attractions as enhancing their lives or attracting tourists (e.g. less interest in festivals, amusement parks, museums, more restaurants);
- see over-development and pollution as the top two concerns about tourism growth, compared to year-round residents who consider higher taxes and traffic congestion as the top two concerns;
- See in larger numbers see a more healthy economy as the number one benefit to having tourism in the community.

One implication is that in some cases, future development that may appear to be agreeable to one segment of the resident population may not be seen in the same way by another. This may necessitate further consideration and processing.

Lakeshore Property Owners as Secondary Stakeholders: 'Our Front Yard'

Lake tourism occurs within the context of either public or private ownership around lake property, or both. The management objective of public (i.e. government) lakeshore property is often well known, as lake areas are typically delineated where public use and access, and facilities are available and appropriate. Private ownership and the needs of lakeshore owners is often less obvious, but equally important. Recent research involving surveys of private owners of lakeshore ownership on five pilot lakes in Minnesota has uncovered some clues to owner sentiment regarding development and lake use (West & Orning 2000). A survey (Arrowhead Regional Development Commission 2003) of nearly 1300 property owners surrounding one large lake (40,087 acres (16,223 ha)) in northern Minnesota revealed:

- 43% reported 'Serious' to 'Very serious' problems of overdevelopment around the lake, with nearly 35% reporting the 'Serious/Very serious' problem to be commercial development;
- nearly one-third said overall water quality was a 'Moderate' problem or worse (e.g. drainage run-off, ineffective septic systems);
- nearly one-third or more reported declining fish, wildlife and aquatic bird habitat as a 'Moderate' problem or worse, and more than 40% reported visual impacts of development seen from the lake as a 'Moderate' problem or worse;
- 31% said the overall water quality was getting 'Worse' over the past five years.

Aside from development impacts, property owners reported other serious problems related to lake use, including use and noise from jet-skis, littering, vandalism and excessive boat-speed problems.

The Commission's study also revealed that lakeshore property owners are responsible for bringing visitors and business to the lake area. On average, lake property owners reported bringing, on average, nearly 16 adults and 9 children-visitors to their property over the past 12-month period. One out of four respondents owned a business and, of those, nearly 10% indicated interest in starting, relocating or expanding their business in the lake destination area. As a group, lakeshore property owners may have more impact on the local economy than first imagined.

This study provided evidence that Minnesota lakeshore owners are very supportive of plans to implement actions to ameliorate problems before they worsen, including using recreational user education, signage and regulations on certain activities for remedy. Beyond environmental and social concerns, they may be highly motivated for other reasons: overdevelopment in lake areas and subsequent water quality degradation has economic implications for themselves and visitors. New research in the headwaters region of the Mississippi River in northern Minnesota has shown a strong relationship between water quality and lakeshore property values (Krysel et al. 2003). Water clarity was shown to be a significant explanatory variable of lakeshore property prices in all regionalised lake groupings involving a total of 37 lakes. Water quality has a positive relationship with property prices. This study, successfully replicating work done on lakes in Maine, concluded that the evidence shows that management of the quality of lakes is important for maintaining the natural and economic assets of the region. Suggesting an area for further research, the implications for the region are that declining water quality could affect the draw (i.e. pull) of a lake destination area and hinder local marketing and development opportunities.

Conclusion

Every lake-destination area will have its own unique characteristics and potential for development. Emerging research and experience in more developed areas points to common themes and needs that lake destination areas should consider. Evidence suggests that pursuing tourism development in these areas that is not sustainable could come at a high cost. A broader scheme that integrates all stakeholders' needs should be employed in value-added development and marketing at the local level.

Research and experience in Minnesota's lake-destination areas suggest the following themes and needs:

(1) Involvement and consideration of all stakeholders are critical to sustainable lake-destination development and marketing.
   Implications: Stakeholders will have differing needs and values. It should be expected that conflicts will arise, but they must be addressed. Actions taken on the front-end of development will avoid costly economic, social and/or environmental development mistakes later that can be difficult or impossible to solve once projects are completed and the tourists arrive. Sustainable tourism development is ultimately a matter of quality customer and stakeholder service.
(2) While lake areas will vary and have unique qualities, the general draw of lakes and water relates to human emotions, or how lakes and water make us *feel* when we are near, as we escape from the varied stresses of our regular day-to-day environment.
   Implications: Marketing campaigns that rely on sound research can help to carefully align the push–pull factors in travel behaviour, focusing on the unique attributes of site-specific destinations and enhancing the chances of hitting the target market. A strong image and brand are integral to marketing success. Attention to customer preferences and needs on-site through visitor-profile research will build and sustain the market, particularly through repeat business.
(3) A strong thread of common environmental sensitivity is woven among all stakeholders in the lake-destination area, as follows:
   - Tourists escaping the stresses of the world and attracted to the qualities that water provides place high priority on high-quality, clean and natural conditions.
   - Providers tend to know what draws their customers (i.e. clean, natural lakes) and they, too, seek to protect those attributes to protect their investments.

- Community and lake-property residents value the natural lake environment they call home and seek to protect those resources against overuse and overdevelopment.
- Environmental protection of the lake destination is also economically and socially motivated.

Implications: Attention to the quality of the lake resource may prove to be the most crucial element to the successful management of image, branding, product development and delivery. Customer and community satisfaction and ultimately business profits, are at stake. Quality marketing and product development can both connect the destination with the consumer for the customer's needs, as well as educate the consumer of the destination's needs – protection of environmental attributes.

(4) Regions and communities interested in developing value-added opportunities for lake-destination tourism will greatly benefit by investing in market research and the development of strategic/master plans the engage all stakeholders in the lake-destination product process.

Implications: Consumer research done well will generate the necessary information to assess markets, understand the client and identify potential new niche markets and value-added development opportunities. Similar processes and background assessments will build strength in businesses and host communities. Sound planning will present the strategic direction and help draw stakeholders together. Issues and value-added options related to such factors as lake access, overnight accommodation development, infrastructure needs and attraction development can all be handled in a comprehensive way that protects social values and resources.

Value-added development opportunities for lake destinations based on sound research and in agreement with stakeholder considerations will likely tend to move towards improving access to lakes and retaining the visitor through broader lodging options and amenity enhancements. Expansion of unique high-end retail opportunities, attractions and nature-based activity opportunities are also viable options. All will be dependent on and driven by, what market research suggests.

Successful destinations will be those that establish principles for future growth and development. Research suggests that such principles need to be oriented towards sustaining the attributes and endowments of lakes and natural areas, including maintaining a region's unique qualities. Guiding principles will contribute to keeping a focus on knowing the customer, provider and community stakeholders and their needs. Time invested in knowing the customer, attending to area tourism businesses so that they can grow and thrive, while at the same time opening conversations with the host communities and lakeshore property owners, will go a long way towards establishing and maintaining a quality experience and sustainability of the destination.

References

Arrowhead Regional Development Centre (2003) *Lake Vermilion survey results: Draft report*. Duluth, Minnesota, MN.

Colle & McVoy (2003) Minnesota Office of Tourism – initial pitch. Working notes. Minneapolis, MN.

Dann, G.M.S. (1977) Anomie, ego-enhancement and tourism. *Annals of Tourism Research* 4 (4),184–94.

Erkkilä, D.L. (1995) Enhancing competitiveness through tourism business retention and expansion programs. In E. Kaynak and T. Erem (eds) *Innovation, Technology and Information Management for Global Development and Competitiveness*. Proceedings of the Fourth Annual World Business Congress, International Management Development Association, 13–16 July. Istanbul, Turkey.

Erkkilä, D.L. and Morse, G.W. (1995) *Itasca County Tourism Business Retention and Expansion Strategies Programme 1995: Executive Summary*. St Paul, MN: University of Minnesota Extension Service, Tourism Centre.

Gursoy, D., Jurowski, C. and Uysal, M. (2002) Resident attitudes: A structural modeling approach. *Annals of Tourism Research* 29 (1), 79–105.

Jurowski, C. (1994) The interplay of elements affecting host community resident attitudes towards tourism: A path analytic approach. Unpublished doctoral dissertation. Blacksburg, VA: Virginia Polytechnic Institute and State University.

Klenosky, D.B. (2002) The 'pull' of tourism destinations: A means-end investigation. *Journal of Travel Research* 40 (4), 385–95.

Krysel, C., Boyer, E.M. Parson, C. and Welle, P. (2003) *Lakeshore Property Values and water quality: Evidence from Property Sales in the Mississippi Headwaters Region*. Report submitted to the Legislative Commission on Minnesota Resources. Brainerd: Mississippi Headwaters Board and Bemidji State University.

Minnesota Office of Tourism (2002) *Tourism Works for Minnesota 2003*. St Paul, MN: Minnesota Office of Tourism.

Norman, W.C., Hamilton, S. and Robertson, S. (1995) *1994 Lac du Flambeau Area Tourism Marketing Research Study*. Madison: Tourism Research and Resource Centre. University of Wisconsin-Cooperative Extension.

University of Minnesota Tourism Centre (2001a) *Community Tourism Development*. Publication No. MI-07650-S. St Paul, MN: University of Minnesota Extension Service.

University of Minnesota Tourism Centre (2001b) *Study of Current Area Tourists: Customer Profiles*. Final report prepared for the Minnesota Office of Tourism. December. St. Paul, MN: University of Minnesota Extension Service.

University of Minnesota Tourism Centre (2002) *Evaluation of the Tourism Market Development Potential of the Itasca Area*. St. Paul, MN: University of Minnesota Extension Service.

West, P. and Orning, G. (2000) Sustainable lakes project: A lake management model for the future. *CURA Reporter*, December. Minneapolis, MN: Centre for Urban and Regional Affairs, University of Minnesota.

Chapter 13
Research Agendas and Issues in Lake Tourism: From Local to Global Concerns

C. MICHAEL HALL AND TUIJA HÄRKÖNEN

As this volume has demonstrated, lakes are a significant dimension of tourism in a number of destinations. The specific environmental characteristics of lakes means that it is appropriate to identify the concept of lake tourism in the same way that the environmental characteristics of mountains means that it is appropriate to talk of alpine tourism. Lake environments are integral to successful lake tourism as lake quality is critical in determining the attractiveness of lakes as destination resources and their use (Härkönen 2003). The overall relationship between tourism and recreation and lake quality is clearly an issue that requires further research, given the relative lack of specific research in this area (see Hall and Härkönen, Chapter 1, this volume). However, there is an awareness of the implications of the impacts of human activities in lake watersheds on lake quality although the regulatory and political will to prevent such activities that negatively affect the lacustrine environment is often lacking. Arguably, it is possible that tourism may have an important role to play as part of a broader economic argument for the maintenance of quality and the changing of behaviours in catchment areas, although broader environmental and human health issues may prove to be more significant. Nevertheless, the importance of the lake environment for determining the sustainability of lake tourism-related businesses may mean that tourism businesses will become key advocates for improved lacustrine environmental management strategies.

Arguably one of the most critical issues in lake management, particularly in increasingly popular destinations that are also subject to amenity-related urbanisation, is to achieve a relative balance between established local communities and amenity migrations, such as second homes and retirement migrants (Hall & Müller 2004). Nevertheless, for more peripheral lake locations such development may prove to be a significant element in economic development strategies. It is only where there is a shortage of housing stock and/or where new development reduces accessibility to lake resources, that conflict may occur with respect to such development. In many cases it is the spirit of place, to use the concept of Tuohino in Chapter 6, related to lakes and the associated amenity values that draws people to locate to such areas. Indeed, an important issue that requires further research is the extent to which lake marketing strategies serve not only to attract tourists but, in the longer term, also influence migration decisions (also see Härkönen 2003).

Although environmental considerations are clearly a limiting factor in the potential development of lake tourism, other issues also emerge. As with many destinations, lake tourism locations are likely to be subject to substantial seasonal variations in use and activities. This is particularly the case for most of the lake destinations discussed in this book – for example, with respect to Finland, the Netherlands, Poland and the United States in which there exists clear summer and winter seasons. Many temperate lake destinations that have cold enough winters to seek to boost winter tourism opportunities through such activities as ice fishing, curling, skating and cross-country skiing. In such destinations, substantial seasonal variation in water temperature may therefore actually be an advantage in comparison with temperate locations where the winter temperatures are not low enough for lakes to freeze. Larger lake tourism destinations also utilise non-lake related tourism, such as events and casinos, as a means to try to extend their season and attractiveness in non-peak periods. However, in some cases seasonality may be related as much to the availability of target fish species and the regulation of fishing and hunting opportunities, as it is to any climatic or other seasonal factor. A further complication for a number of lake destinations is that their location may be relatively peripheral in relation to tourism generating regions. Again, this is an issue for much lake tourism in Nordic countries and North America. This is a situation that may exacerbate problems of seasonality and competition with other destinations, as well as limit the potential growth of lake tourism, although, arguably, this limit on tourist movement may also serve to help maintain some of the environmental qualities of lakes in some situations. Nevertheless, the seasonal nature of much water-based recreational and tourism activities does create significant pressures on businesses in lake tourism locations in terms of the development of business strategies that manage seasonality, particularly if there are limited business opportunities in the low demand period (Hall & Boyd 2005). Therefore, a further area of research is the extent to which lake tourism businesses demonstrate strategies that are differentiated from those of businesses in other locations. This would include information with respect to the nature of the entrepreneurship of lake tourism firms, including the extent to which they demonstrate the characteristics of being lifestyle businesses, their networking strategies and the extent to which they adopt strong environmental positions, given their dependency on lake quality (also see Härkönen 2003; Erkkilä, Chapter 12, this volume).

An increasing issue of many lake destinations is the issue of accessibility and rights of use. Although there are clear differences in access rights in Nordic countries compared to New Zealand and North America, issues of access and use are becoming increasingly contested as pressures on scarce water resources grow. Lakes obviously have great potential to meet different tourists' needs, be it active participation, such as kayaking, or passive observation, such as admiring lake scenery. However, different needs will have different access and use requirements and, at times, such use requirements may be at odds with each other, as well as those of other users of lake and water resources. Indeed, use issues will arguably be a major area of lake management research in years to come. It should also be noted that the majority of research, as well as the focus of this book, has been on tourism in temperate lake environments in the developed world. There is clearly also a need to examine the management and tourism dimensions of lacustrine systems in other environmental and economic conditions, including not only in the tropics, but also lakes, including seasonal and 'dry' lakes in arid environments. For example, Lake Deborah in Western Australia is a salt lake system surrounded by arid shrublands that fills about once every ten years (Department of the Environment 2003). Such extreme habitats with their unique biodiversity and conditions may prove attractive to some tourists, as well as being intrinsically valued for their environment. In addition, future research may also serve to identify the different environmental, management and use issues that arise between natural lakes and artificial reservoirs and dams. Again, using an example from Western Australia, the Ord River Dam which created Lake Argylle,

> changed the flow in the Ord River from seasonal large flows, to less intense, continual flows resulting in ecological changes and impacts on cultural values. The impacts on our ecological, cultural, aesthetic, recreational and economic values of our waterways are enormous but often go unnoticed until the function or value of a waterway is compromised, such as by algal blooms. (Department of the Environment 2003)

Although, as the various chapters in this volume indicate (especially Chapters 1 and 10–12), there is no 'one size fits all' management approach for all lacustrine tourism systems, there are a number of common elements that need to be identified (see Figure 1.2). Yet there are likely to be a number of topics that deserve further research in their own right, as well as in terms of their interaction with each other. These include evalation, research and monitoring; the development of action plans on a spatial basis such as a watershed or catchment; funding arrangements; implementation strategies; policy development; statutory and regulatory planning procedures; education, communication and interpretation; and the development of cooperative relations between stakeholders, including tourism and non-tourism stakeholders. Indeed, attention to management approaches is arguably becoming all the more important, given the increased stress that is being placed on lakes by human activities, many of which have little to do with tourism, although possibly the most significant impact on many lakes will be the impact of global environmental change.

Lakes and Global Environmental Change

Issues of global environmental change (GEC) also pose significant challenges for many lake destinations (Hulme et al. 2003; Håkanson et al. 2003; Hall & Higham 2005; Gössling & Hall 2005; Jones et al. 2005). Climate change is obviously one very significant element of GEC that has the potential to impact on many lacustrine systems (e.g. Jones et al. 2005) and this will be discussed in more detail below. However, another significant dimension of GEC are issues of biodiversity and the spread of diseases and pests. In some instances, lake habitat has been deliberately altered in an effort to make it attractive for a more desirable fish species over other species in order to enhance recreational fishing opportunities (Wills et al. 2004), although more typical is the introduction of such desirable species into an existing habitat in which it is believed they will thrive. For example, there is a long history of fish species being deliberately introduced from one location to another outside their previous range in order to enhance recreational fishing opportunities (Huckins et al. 2000). In the case of the United States, the stocking of wilderness lakes with trout began in the 1800s (Pister 2001). This practice was followed for nearly a century with the singular goal of creating and enhancing sport fishing, and without any consideration of the ecological ramifications. Changes to practices only started to occur in the 1960s when research indicated negative impacts on the native biota which were attributable to introduced species. As Pister (2001) notes, the necessity for wilderness fish stocking is now the subject of widespread debate in the United States, as in other countries such as Australia and New Zealand, where trout were introduced for the purposes of sport fishing and the Europeanisation of the environment, especially in view of changing social values and priorities with respect to preserving the biodiversity of mountain lake ecosystems. Indeed, in some locations the stocking of lakes with fish for recreational purposes has had significant unexpected impacts. For example, in the case of Lake Veere in southwest Netherlands, although the growth and initial production of the introduced trout population has been attractive from the perspective of a recreational fishery, the high mortality rate and infestation of trout with the salmon louse has become a serious drawback for a future stocking programme (Raat 2003). Nevertheless, while the deliberate introduction of species is a significant point of debate between different stakeholders in lake-based tourism and recreation because of the trade-offs between ecological impacts and recreational benefits (Huckins et al. 2000), there is typically little debate over the damage caused by the unplanned introductions of species.

One plant pest that has typically been accidentally spread by recreationists is Eurasian water milfoil (*Myriophyllum spicatum*), an aquatic invasive weed that has been identified recently at a number of sites in the western United States, including Lake Tahoe. Because Eurasian water milfoil is easily spread by fragments, transport on boats and boating equipment plays a key role in contaminating new water bodies. Unless the weed is controlled, significant alterations of aquatic ecosystems, with its associated degradation of natural resources and economic damage to human uses of those resources, particularly tourism and recreation, may occur. In a study of the value of a portion of the recreational service flows that society currently enjoys in the Truckee river watershed below Lake Tahoe, Eiswerth et al. (2000) calculated that the lower-bound estimates of baseline water-based recreation value at a subset of sites in the watershed ranged from US$30 million to US$45 million per annum. Given such economic significance of water-based recreation, the impacts from the continued spread of Eurasian water milfoil in the watershed could be extremely significant and they suggested that even a 1% decrease in recreational values would correspond to losses of approximately US$500,000 per annum as a lower bound estimate.

Another pest species that is being accidentally dispersed by recreational boaters in the United States is the zebra mussel, which is an extremely invasive bivalve that displaces native species. In a study conducted on Lake St Clair in Michigan, Johnson et al. (2001) identified several mechanisms associated with recreational boating that were capable of transporting either larval or adult life stages of the mussel. Larvae were found in all forms of water carried by boats (i.e. in live wells, bilges, bait buckets and engines), but were estimated to be 40–100 times more abundant in live wells than in other locations. They also noted that, contrary to common belief, mussel dispersal from these boat launches did not occur by direct attachment to transient boats. Instead, adult and juvenile mussels were transported primarily on macrophytes entangled on boat trailers (5.3% of departing boats inspected) and on anchors (0.9% of departing boats). By combining these data with estimates of survival in air and reported boater destinations, Johnson et al. (2001) predicted that a maximum of 0.12% of the trailered boats departing the access sites at which inspections were conducted delivered live adult mussels to inland waters solely by transport on entangled macrophytes. Although they noted that this was a small probability, the high levels of boating activity resulted in a prediction of a total of 170 dispersal events to inland waters within the summer season from the primary boat launch studied.

The invasion of lake systems by exotic fauna and flora as a result of tourism and recreational activity arguably becomes even more problematic because of the potential effects of climate change. Climate change can have significant impacts on hydrological systems (Mortsch et al. 2000) and other biotic and abiotic systems with which the hydrological system interacts. Even seemingly minor changes in water and air temperature – micro-, meso- and macro-climate and weather patterns – can have a substantial effect on habitat and on the range of flora and fauna. Climate change can lead to environmental stress on existing species and provide opportunities for new species, although it must be emphasised that species' response to climate change is not homogeneous (Levin 2003). In North America, it is predicted that the range of cool-water fish (e.g. wall-eye, perch) and particularly warm-water fish species (e.g. bass), will expand northward as a result of increased water temperatures and this will alter the composition of recreational fish catches (Jones et al. 2005). A study by the US Environmental Protection Agency (1995) suggested that the thermal habitat for many cold-water sport fish (e.g. rainbow, brook and brown trout) would be reduced by 50 to 100% in the Great Lakes region under a doubling of atmospheric carbon dioxide (~2050s). Under similar climate change scenarios, other studies have shown that warmer water temperatures would eliminate the sport trout fishery from most North Carolina streams (Ahn et al. 2000) and reduce the habitat for several popular cool-water sport fish in eastern Canada, including wall-eye, northern pike and white fish (Minns & Moore 1992).

However, arguably the direct effects of climate change on the lacustrine environment will be compounded by the interaction of climate change with other human impacts in lake watersheds, with a consequent effect on lake water quality, including the solubility of dissolved oxygen, the metabolism and respiration of plants and animals and the toxicity of pollutants (Stefan et al. 1998) which, as several chapters in this volume have indicated, can have a significant impact on tourist perceptions of attractiveness. Furthermore, climate change in tandem with land-use can also impact the run-off of nutrients into lake systems, leading to issues of eutrophication and littoral algal production (Cronberg 1999). In severe cases eutrophication can have a substantial impact on tourist and recreational activities, not only because of the impact of recreational fishing, but also because of issues associated with the smell of algae washed up on to the shoreline, as well as the potential for toxic algal blooms that can make the water unsafe to swim in.

Another potential impact of climate change on some lake systems where inflows and water levels are reduced is not only the direct impacts of such changes on recreational activities, such as boating, fishing and swimming, but the increased competition between different users for an increasingly limited supply of water. For example, trade-offs in some locations may be sought between water for golf courses versus water for agriculture or even drinking. Indeed, tourism development may be constrained if there is not adequate water supplies (Gössling 2001; Kent et al. 2002; Essex et al. 2004). As Gössling (2005) has argued, tourism production can exacerbate freshwater problems, as it is often concentrated in regions with limited water resources. According to Gössling (2002, 2005), besides causing a shift in global water consumption from regions of relative water abundance to those that are water scarce, tourism also increases total water demand as people use larger quantities of this resource when they are on vacation.

Fresh water is becoming an increasingly valuable commodity whether for direct use or amenity (e.g. Raphael & Javonski 1979; Farber & Costanza 1987; Oglethorpe & Miliadou 2000; Goetgeluk et al. 2005) and a growing area of interest in terms of systems of governance (Moore 2003; Bruch 2004; Uitto 2004). Global environmental change and climate change and population growth in particular (Vörösmarty et al. 2000; Arnell 2004), are placing increased stress on freshwater sources, including lake systems. Indeed, one of the greatest threats to the water quality of lacustrine systems as well as the lakes themselves is land-use in lake watersheds. In many lake systems, deforestation in headwaters is leading to accelerated erosion in producing sediments that degrade the quality of water (Haigh et al. 2004). The effects of deforestation may be further compounded by inappropriate agricultural practices, as well as poorly managed urban development.

Indeed, not only are environmental changes becoming global in scope, but it is increasingly recognised that management responses will also often need to have an international component. As Uitto (2004: 5) notes:

> Competition over limited water resources may increase the risk of conflict, especially when there are already other sources of tension between the riparian countries. It is, therefore, essential to develop effective international mechanisms for the governance of shared water bodies.

Such a situation clearly exists with issues involved in the management of transborder lake environmental issues (Bruch 2004), as well as transborder lake tourism (Leimgruber 1998) with respect to international regulatory regimes. Examples of international conventions that affect lake management include the 1992 *UN/ECE* Convention on the Protection and Use of Transboundary Watercourses and International Lakes, the Ramsar Convention on Wetlands of International Importance (1971), the UN Convention on Non-navigational Uses of International Watercourses (1997), the Bellagio Draft Agreement Concerning the Use of Transboundary Groundwaters (1987) (Bruch 2004; Uitto 2004). In addition, a number of supranational agreements exist with respect to lakes such as the 1995 Protocol on Shared Watercourse Systems in the Southern African Development Community and transboundary agreements and authorities such as those for the Great Lakes and Lake Victoria (Bruch 2004). Although there is no comprehensive list of the number of lakes subject to more that one national jurisdiction (what might be described as international lakes), Uitto (2004) makes reference to 214 international river basins, with the final number being relatively fluid because of changes in national status. However, with increasing realisation of the interconnectedness of environmental systems and of human influence on such systems, it is readily apparent that lakes are a significant part of a broader awareness of the implications of global environmental change and that the potential impacts of humans in one part of the planet can lead to environmental changes that can dramatically impact lacustrine systems on the other side of the world.

Conclusion

The human impacts on lacustrine systems can be dramatic. The Aral Sea has almost vanished as a result of human interference in the watershed, the marine biodiversity of the Great Lakes has changed dramatically as a result of the introduction of exotic species and pollution and Lake Toolibin near Narrogin is the only freshwater lake left in the wheat-belt of Western Australia (Department of the Environment 2003) with other freshwater lakes having become saline as a result of land clearance and climate change. Clearly, such circumstances have dramatic implications not only for the lake environment but also for recreational and tourist use and attractiveness. As noted above, the condition of lakes and waterways has now become a global, as well as a local concern and have become recognised as an essential element of sustainable development (e.g. Klessig 2001). Yet, arguably, in the same way that the broader tourism industry and particularly international organisations such as the World Tourism Organisation and the World Travel and Tourism Council has embraced sustainable tourism development rather than sustainable development – which of course considers tourism as just one element of a range of sustainable development strategies – so it is that there is a danger that lake tourism will be seen as an end in itself rather than as a means. Tourism, as with all human activities, is ultimately dependent on the environment. In developing lake tourism it therefore becomes paramount that an integrated approach to lacustrine systems is adopted that acknowledges the role of tourism as just one component of the complex management mix that surrounds lakes. It is likely that where this is done, the broader economic and social well-being, as well as, of course, environmental well-being will be accomplished. Yet to achieve this not only requires an improvement in the state of knowledge about lake tourism in the tourism industry, but also other users and managers of lakes as well as the communities that depend on them and the policy-makers that ultimately bear responsibility for the political decisions that surround lacustrine systems. We hope that this book is one small foot in the water towards achieving this goal.

References

Ahn, S., De Steiguer, J., Palmquest, R. and Holmes, T. (2000) Economic analysis of the potential impact of climate change on recreational trout fishing in the southern Appalachians: An application of a nested multinomial logit model. *Climatic Change* 45, 493–509.

Arnell, N.W. (2004) Climate change and global water resources: SRES emissions and socio-economic scenarios. *Global Environmental Change* 14, 31–52.

Bruch, C.E. (2004) New tools for governing international watercourses. *Global Environmental Change* 14, 15–23.

Cronberg, G. (1999) Qualitative and quantitative investigations of phytoplankton in Lake Ringsjön, Scania, Sweden. *Hydrobiologia* 404, 27–40.

Department of the Environment (2003) *Western Australia's Waterways: Understand, Protect, Restore*. Perth: Department of the Environment.

Eiswerth, M.E., Donaldson, S.G. and Johnson, W.S. (2000) Potential environmental impacts and economic damages of Eurasian watermilfoil (*Myriophyllum spicatum*) in western Nevada and northeastern California. *Weed Technology* 14 (3), 511–18.

Essex, S., Kent, M. and Newnham, R. (2004) Tourism development in Mallorca. Is water supply a constraint? *Journal of Sustainable Tourism* 12 (1), 4–28.

Farber, S.C. and Costanza, R. (1987) The economic value of wetland systems. *Journal of Environmental Management* 24, 41–51.

Goetgeluk, R., Kauko, T. and Priemus, H. (2005) Can red pay for blue? Methods to estimate the added value of water in residential environments. *Journal of Environmental Planning and Management* 48 (1), 103–20.

Gössling, S. (2001) The consequences of tourism for sustainable water use on a tropical island: Zanzibar, Tanzania. *Journal of Environmental Management* 61 (2), 179–91.

Gössling, S. (2002) Global environmental consequences of tourism. *Global Environmental Change* 12 (4), 283–302.

Gössling, S. (2005) Tourism and water. In S. Gössling and C.M. Hall (eds) *Tourism and Global Environmental Change*. London: Routledge.

Gössling, S. and Hall, C.M. (eds) (2005) *Tourism and Global Environmental Change*. London: Routledge.

Haigh, M.J., Lansky, L. and Hellin, J. (2004) Headwater deforestation: A challenge for environmental management. *Global Environmental Change* 14, 51–61.

Håkanson, L., Ostapenia, A., Parparov, A., Hambright, K.D. and Boulion, V.V. (2003) Management criteria for lake ecosystems applied to case studies of changes in nutrient loading and climate change. *Lakes & Reservoirs: Research and Management* 8, 141–55.

Hall, C.M. and Boyd, S. (eds) (2005) *Nature-based Tourism in Peripheral Areas: Development or Disaster.* Clevedon: Channel View Publications.

Hall, C.M. and Higham, J. (eds) (2005) *Tourism, Recreation and Climate Change.* Clevedon: Channel View Publications.

Hall, C.M. and Müller, D. (eds) (2004) *Tourism, Mobility and Second Homes: Between Elite Landscape and Common Ground.* Clevedon: Channel View Publications.

Härkönen, T. (ed.) (2003) *International Lake Tourism Conference, 2–5 July, 2003, Savonlinna, Finland.* Savonlinna: Savonlinna Institute for Regional Development and Research, University of Joensuu.

Huckins, C.J.F., Osenberg, C.W. and Mittelbach, G.G. (2000) Species introductions and their ecological consequences: An example with congeneric sunfish. *Ecological Applications* 10 (2), 612–25.

Hulme, M., Conway, D. and Lu, X. (2003) *Climate Change: An Overview and its Impact on the Living Lakes.* A report prepared for the 8 Living Lakes Conference, Climate change and governance: managing impacts on lakes, held at the Zuckerman Institute for Connective Environmental Research, University of East Anglia, Norwich, UK, 7–12 September. Norwich: Tyndall Centre for Climate Change Research.

Johnson, L.E., Ricciardi, A. and Carlton, J.T. (2001) Overland dispersal of aquatic invasive species: A risk assessment of transient recreational boating. *Ecological Applications* 11 (6), 1789–99.

Jones, B., Scott, D. and Gössling, S. (2005) Lakes and streams. In S. Gössling and C.M. Hall (eds) *Tourism and Global Environmental Change.* London: Routledge.

Kent, M., Newnham, R. and Essex, S. (2002) Tourism and sustainable water supply in Mallorca: A geographical analysis. *Applied Geography* 22, 351–74.

Klessig, L.L. (2001) Lakes and society: The contribution of lakes to sustainable societies. *Lakes and Reservoirs: Research and Management* 6 (1), 95–101.

Leimgruber, W. (1998) Defying political boundaries: Transborder tourism in a regional context. *Visions in Leisure and Business* 17 (3), 8–29.

Levin, P.S. (2003) Regional differences in responses of Chinook salmon populations to large-scale climatic patterns. *Journal of Biogeography* 30, 711–17.

Minns, C. and Moore, J. (1992) Predicting the impact of climate change on the spatial pattern of freshwater fish yield capability in eastern Canadian lakes. *Climatic Change* 22, 327–46.

Moore, A.O. (2003) *Living Lakes Governance: What the Survey Shows and Next Steps.* Presentation prepared for 8th Living Lakes Conference, Norwich, England, 11 September.

Mortsch, L., Hengeveld, H., Lister, M., Logfren, B., Quinn, F., Slivitzky, M. and Wenger, L. (2000) Climate change impacts on the hydrology of the Great Lakes – St Lawrence system. *Canadian Water Resources Association Journal* 25 (2), 153–79.

Oglethorpe, D.R. and Miliadou, D. (2000) Economic valuation of the non-use attributes of a wetland: A case-study for Lake Kerkini. *Journal of Environmental Planning and Management* 43 (6), 755–67.

Pister, E.P. (2001) Wilderness fish stocking: History and perspective. *Ecosystems* 4 (4), 279–86.

Raat, A.J.P. (2003) Stocking of sea trout, *Salmo trutta*, in Lake Veere, south-west Netherlands. *Fisheries Management and Ecology* 10 (2), 61–71.

Raphael, C.N. and Jawonski, E. (1979) Economic value of fish, wildlife and recreation in Michigan's coastal wetlands. *Coastal Zone Management Journal* 5, 181–94.

Stefan, H.G., Fang, X. and Hondzo, M. (1998) Simulated climate change effects on year-round water temperatures in temperate zone lakes. *Climatic Change* 40, 547–76.

Uitto, J.I. (2004) Multi-country cooperation around shared waters: Role of monitoring and evaluation. *Global Environmental Change* 14, 5–14.

United States Environmental Protection Agency (1995) *Ecological Impacts from Climate Change: An Economic Analysis of Freshwater Recreational Fishing*, Report No. 220-R-95–004, Washington, DC: US Environmental Protection Agency.

Vörösmarty, C.J., Green, P., Salisbury, J. and Lammers, R.B. (2000) Global water resources: Vulnerability from climate change and population growth. *Science* 289, 284–8.

Wills, T.C., Bremigan, M.T. and Hayes, D.B. (2004) Variable effects of habitat enhancement structures across species and habitats in Michigan reservoirs. *Transactions of the American Fisheries Society* 133 (2), 399–411.

Index

activity choice 121, 150-162, 184-5
Aral Sea 30, 230
Australia 7, 17, 225, 230
– New South Wales 17
– Western Australia 225, 230

biomes 3-4
boating 5, 47, 121-30, 141-4, 149-64

caldera lakes 45-6
Canada 39
canals 131-8
climate change 228
collaboration 16-8, 30, 84, 199
consumption centres 169-71
Crater Lake 45-66
cultural values 28, 47, 50-8, 67-80, 83-6, 105-6

destinations 28

economic values 5, 127-8. 173-8, 208
environmental issues 7-9, 14, 15, 31-2, 63-64, 189-90, 226-30
environmental values 7, 45-66, 77-8
Eurasian water milfoil 227

Finland 67-82, 101-18, 149-64
– Savonlinna 70-82
fishing 9-10, 108, 112-3, 226-7
Fraser Island 7, 31

gateways 171-2, 179
global environmental change 226-30
Great Lakes (North America) 5, 9, 39
Great Lakes United 30

historical dimensions 45-66, 67-82
hydroelectric development 196-7

icons 51-8, 68, 182
imagery 56-8, 67-82, 108-18, 211-3
indigenous peoples 54, 83-97
institutional arrangements 14-6

integrated lacustrine management 11-8, 30, 34-40
integrated lake management, *see* integrated lacustrine management
interpretation 86-7
Israel 9

Japan 11-2

Lake Argyle 225
Lake Biwa 11-2, 31
Lake Deborah 225
lake destination attributes 28-40, 207-8
Lake Dongting Hu 31
Lake District (UK) 6, 37-9
lake environments 7-11, 27-8
Lake Kinneret 8
Lake Mälaren 8
Lake Pukaki 91-4
Lake St Clair 227
Lake Tahoe 227
Lake Taupo 88-9, 189, 190-91
Lake Te Anau 89-91
Lake Titicaca 29, 32
Lake Toolibin 230
lake tourism, significance of 4-7
Lake Wakatipu 195
Lake Windermere 37-9
landscape perception and representation 6-7, 50-64, 67-82, 101-5

marketing 101-18, 129-30, 207-22
multiple use 32-34, *see also* integrated lacustrine management

national landscapes 67-9, 71-4
national parks 3-4, 6, 37-9, 45-66, 89
Netherlands 119-30
New Zealand 8, 83-97, 182-206

Peru 29
pest species 227-8
pleasure craft, *see* boating

Index

Poland 131-48
pollution 7-11, 31; *see also* water quality
planning 37-40, 180, 219-22, *see also* integrated lacustrine management

recreation opportunity spectrum (ROS) 34
reservoirs 33, 225
Romantic movement 6, *see also* sublime
Rotorua Lakes 192-3
rural tourism 168-9

seasonality 224
sense of place 87, 101-5
setback regulations 37
shoreline management 31-2
steamboat 74-5
sublime 50-8, *see also* Romantic movement
sustainable management 194-203, 230
Sweden 8-9
swimming 7

topophilia 103
topophobia 103
tourism development 144-7, 167-8, 178-9, 215-22, 223

tourism marketing 79, 101-18, 207-8
Tourism Pressure Index 31
tourism promotion 56-63, 144-7, 182
transborder lake management 229-30
trip motivation 209-16

United Kingdom 6. 37-9
United States of America 10, 39, 45-66, 167-81, 227
– Iowa 9
– Michigan 227
– Minnesota 9, 172-81, 208-22
– Oregon 17, 45-66

urbanisation 7, 11-2
user competition and conflict 33-4, 194, 219-22, 223, 225, 229-30

water catchments 11-3, 16, 17, 64
water quality 7-9, 14, 15, 64, 189-90
water-borne disease 10
watershed management, *see* integrated lacustrine management
wilderness 63-4
Wisconsin 30

zebra mussel 227-8